T0176830

# PRACTICAL LIGHTING DESIGN WITH LEDs

SECOND EDITION

# PRACTICAL LIGHTING DESIGN WITH LEDs

RON LENK

CAROL LENK

Mohamed E. El-Hawary, *Series Editor*

Published by John Wiley & Sons, Inc., Hoboken, New Jersey.
Published simultaneously in Canada.

For general information on our other products and services or for technical support, please contact our Customer Care Department within the United States at (800) 762-2974, outside the United States at (317) 572-3993 or fax (317) 572-4002.

Wiley also publishes its books in a variety of electronic formats. Some content that appears in print may not be available in electronic formats. For more information about Wiley products, visit our web site at www.wiley.com.

*Library of Congress Cataloging-in-Publication Data is available.*

ISBN: 978-1-119-16531-6

Printed in the United States of America.

10  9  8  7  6  5  4  3  2  1

*To our children, for being so patient*

# CONTENTS

# ABOUT THE AUTHORS

**Ron Lenk** is an authority in the fields of power electronics, power systems and LED drivers. The author of the bestselling *Practical Design of Power Supplies* (Wiley), he has spent the last twelve years working on LEDs and lighting. Lenk co-founded and was CEO at Switch Light, Inc. which made general-service LED light bulbs, and now is a consultant in the fields of power and LEDs. He is a Senior Member of the IEEE and has 35 issued US patents.

**Carol Lenk** was the co-founder and Director of Engineering at Switch Light, Inc. She earned a B.S. in electrical engineering from MIT and a master's in math and science education. One of the pioneers in applying LEDs to general lighting, Lenk has ten years' experience combining theoretical concepts with practical engineering in fields as diverse as optics, thermal modeling, material science, electronics and mechanical design. She is now a consultant in the field of LEDs and has nine issued US patents, all relating to LED lighting.

# *PREFACE*

## THE LIGHTING REVOLUTION

LEDs are bringing in a new era in lighting. Similar to the evolution of computing power that computers went through, from vacuum tubes to the silicon-based semiconductor brains of modern-day computers, lighting is now riding an exponential growth wave in efficacy. From oil lamps to the invention of the Edison light bulb 100 years ago to the fluorescent lights of 50 years ago to the LEDs of today, lighting technology is finally joining the modern world of solid-state technology.

LED-based lighting is increasingly becoming the efficient light source of choice, replacing both incandescent and fluorescents. The hurdles that have kept consumers from adopting energy-efficient lighting, such as shape, color quality, the presence of toxic mercury, and limited lifetime, are all better addressed by LEDs.

In the long term, LED-based lighting will be better and cheaper than every other light source. It will become the *de facto* light of choice. LED lighting will be cheap, efficient, and used in ways that haven't been imagined yet. It will transform the $100 billion lighting industry, and with transformation comes opportunity.

## THE LAST VACUUM TUBE

Lighting is the last field that still uses vacuum tubes. All electronics today use integrated circuits because of the enormous benefits in performance and cost. But a fluorescent tube is a type of vacuum tube. LEDs are solid-state devices, the same as the rest of electronics. The amount of light that an LED can convert from 1 W of power is already nearly double that of the best fluorescent tubes. The future is even brighter as LEDs are anticipated to continue that growth in the next decade, and then soon go on to reach the physical limits of electricity to light conversion. We look forward to seeing the last ceiling-mounted vacuum tube in the not-too-distant future.

## GREEN LIGHTING

The benefits of using LEDs for lighting are many. The most obvious is their efficiency. Lighting accounts for 20% of total electricity use throughout the world today. Using LEDs could cut this down to 4% or less. As LEDs become the dominant light source over the next decade, the reduction of energy used and greenhouse gases emitted will benefit everyone. Each consumer will save hundreds of dollars every year from reduced energy use. Building owners will save even more. Utilities will be better equipped to manage growth. And the earth will experience the accumulation of fewer greenhouse gases, as well as a reduction in the emission of toxic mercury found in fluorescent lighting.

## A LIFETIME OF LIGHTING

As solid-state devices, LEDs have extremely long lifetimes. They have no filaments to break. They can't leak air into their vacuum because they don't use a vacuum. In fact, they don't really break at all; they just very gradually get dimmer. Imagine changing your light bulb only once or twice in your entire lifetime!

## LIGHTING THE WORLD WITH LEDS

Just as microprocessors got cheaper and more powerful, LEDs are also benefiting from the cost-reduction techniques developed in the semiconductor industry. LED light prices are already on par with fluorescent tubes. And with lower prices will come the ability to tailor light to the specific needs of the consumer. Taken together with LEDs' reduced energy usage, this will enable the universal availability of lighting. Imagine every child in the poorest village having a light to read by.

The design of LED-based lighting systems is an exciting field, but these systems are fairly technical. With this book, we hope to enable the readers to do great things with lighting, both for themselves and for the world.

*Marietta, Georgia*                                                                RON LENK
                                                                                 CAROL LENK

# LIST OF FIGURES

CHAPTER *1*

# PRACTICAL INTRODUCTION TO LEDs

LIGHT BULBS are everywhere. There are over 20 *billion* light bulbs in use around the world today. That is, three for each person on the planet! We expect that within the next 10 years, the majority of these bulbs will be light-emitting diodes (LEDs). This is because LEDs can provide efficiency dozens of times higher than incandescent light bulbs. They can be as efficient as the theoretical limit for electricity to light conversion set by physics. This book is all about the practical aspects of LEDs and how you can make practical lighting designs using them.

## WHAT IS AN LED?

The purpose of this book is to tell you practical things about LEDs. So in this section, we're not going to regale you with jargon about "direct bandgap GaInP/GaP strained quantum wells" or such. Let's directly address the question: What is an LED?

The name "light-emitting diode" tells you a lot already. In the first place, the noun tells you that it is a diode. A diode conducts current in one direction and not the other. And that's what an LED does. While we'll explore the details of its electrical behavior in Chapter 4, the only thing to note for the moment is that it has a much higher forward voltage than the diodes usually used in electronics. While a 1N4148 has a drop of about 700 mV, an LED may drop 3.6 V. This is because LEDs are not made from silicon, but from other semiconductors. But other than that, an LED's electrical characteristics are very much like those of other diodes.

The words "light-emitting" tell you a lot more. Now all diodes emit at least a little bit of light. You can open up an integrated circuit (IC) and use a scanner to see which parts of the circuit are emitting light. This tells you which parts are conducting current. IC designers use this to help debug their ICs. However, the amount of light emitted by ICs is very small. Since the purpose of LEDs is to emit light, they have been carefully designed to optimize this performance. That's why, for example, they have a much higher forward voltage than normal, rectifier diodes. Rectifiers have been optimized to minimize their forward voltage while maximizing reverse breakdown

*Practical Lighting Design with LEDs*, Second Edition. Ron Lenk and Carol Lenk.
© 2017 by The Institute of Electrical and Electronics Engineers, Inc. Published 2017 by John Wiley & Sons, Inc.

voltage. LEDs are optimized to produce the most light of the right color at the lowest power, and things such as forward voltage (by itself) don't matter. Of course, forward voltage does enter into how much power the LED dissipates, and we'll see in Chapter 5 how to characterize the light emitted versus the power dissipated.

## SMALL LEDs VERSUS POWER DEVICES

Present-day thinking divides LEDs into two classes: small devices and power devices. Small LEDs became widely used in the 1970s. They come in all different colors, such as red, orange, green, yellow, and blue. They are the small T1¾ (5 mm) devices shown in Figure 1.1. Nowadays, there are literally tens of billions of them sold each year. They go into cell phone backlights, elevator pushbuttons, flashlights, incandescent bulb replacements, fluorescent tube replacements, road signage, truck taillights, traffic lights, automobile dashboards, and so on.

What characterizes these small devices is their power level, or as the industry thinks of it, their drive current. The typical red small LED, for example, has a drive current of 20 mA. At a forward voltage of 2.2 V, this is only 44 mW of power. (The efficacy is so low that this is just about equal to the heat dissipation as well.) Small white LEDs have a higher forward voltage (3.6 V, corresponding to 72 mW), and some small LEDs can be run as high as 100 mA. But fundamentally, this type of LED is used as an indicator, not a real light source. It takes 14 of them to make a somewhat reasonable 1 W flashlight, and hundreds of them to make a (dim) fluorescent tube replacement.

While the information in this book is applicable to these small LEDs, the main focus is on power devices. Power devices are typically 1–3 W devices that are usually run at 350 mA. Their dice (the actual semiconductor, as opposed to its package) are substantially larger than those of small LEDs, although their footprint need not be.

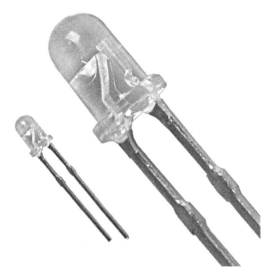

Figure 1.1   T1¾ (5 mm) LEDs.

These devices are typically used in places requiring lighting, rather than as indicators. Applications include flashlights, incandescent bulb replacements, large-screen TVs, projector lights, automotive headlights, airstrip runway lighting, and just about everywhere lighting is used. Of course, not all of these applications have yet seen widespread adoption of power LEDs, but they will soon.

## PHOSPHORS VERSUS RGB

Most lighting designs are going to be made with white light (which includes incandescent "yellow" light). For this reason, this book concentrates primarily on white LEDs. However, what is described here for white LEDs can be straightforwardly applied to color LEDs. Color LEDs are very similar to white, albeit with differing forward voltage. The reason for the varying forward voltages is that the colored light (red, yellow, blue, etc.) is generated directly by the semiconductor material. The material is varied to get differing colors and the differences in material in turn cause differences in forward voltages.

However, white light cannot be directly generated by a single material (we are ignoring special types of engineered materials that are not yet in production). White light consists of a mixture of all of the colors. You already know this because white light can be separated into its constituent colors with a prism. White light thus has to be created. There are currently two main methods of generating white light with LEDs. In one method, an LED that emits blue light is used, and the blue light is converted to white by a phosphor. In the other method, a combination of different color LEDs is used.

The first method is the most common. A typical wavelength for the blue light generated by the LED is 435 nm. Why use blue light? This has to do with the physics of the way the white light is generated. The blue light is absorbed by a phosphor, and re-emitted as a broad spectrum of light approximating white. For the phosphor to be able to absorb and re-emit the light, the light coming out has to be lower in energy than the light going in. That's just like any electronic component. Energy goes in, some is dissipated as heat, and the rest comes out again, transformed. So to get all of the colors in the spectrum that humans can see, the phosphor needs to have input at a higher energy (shorter wavelength) than the shortest color's energy. For humans, this is about 450 nm, and so a 435 nm blue LED is the most energy-efficient way of generating white light using a phosphor.

Before turning to the second method of generating white light, we should say a few more words about the phosphor. There are various types of phosphors. Phosphors are designed to absorb one specific wavelength of light, and re-emit it at either one or more different wavelengths or in a band of wavelengths. LED phosphors are typically designed to do the latter. But there are limits to how broad a band of colors a phosphor can emit. So many LEDs use bi-band or tri-band phosphors to better cover the spectrum of light needed to approximate white. These phosphors are mixtures of two or three primary phosphors. These more complicated phosphors are typically used when better color rendition is needed (see the discussion of color rendering index (CRI) in Chapter 3).

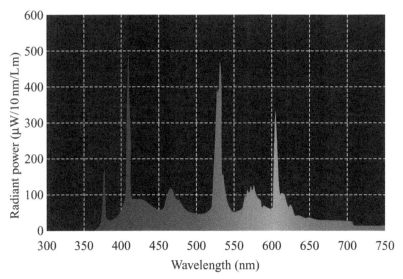

Figure 1.2    Fluorescent tube's spectral power distribution. (*Source:* http://www.gelighting
.com/LightingWeb/na/resources/tools/lamp-and-ballast/pop_curves.jsp?12.) (See color plate
section.)

As a side note, we can comment briefly on fluorescent lights. In some ways, a
fluorescent light is quite similar to an LED, but its fundamental mechanism of light
emission is different. It generates a high-temperature plasma inside a tube, which
emits light in the ultraviolet (UV) range (254 nm) rather than in the blue range. But
after that, it too uses a phosphor to absorb the light and re-emit it in the visible range.
Note that since the wavelength of the light is considerably farther away from the
visible spectrum than the 435 nm generated electrically by the LED die, the efficiency
ultimately possible for a fluorescent is intrinsically lower than that possible for an
LED. (At the moment, fluorescent lights and LEDs have roughly the same efficiency.)

But also interesting is the type of phosphor the typical fluorescent light uses.
These phosphors are of the type that re-emits in just one or two narrow wavelengths,
not in a band of colors. The specific wavelengths emitted have been very carefully
chosen to make the light emitted give a good specification for the CRI. But the spiky
nature of the emission spectrum (see Fig. 1.2) means that colors at wavelengths other
than these are poorly reproduced by the fluorescent lamp. Of course, there is no reason
(we know of) that fluorescents can't have the same spectra that LEDs do. But for the
present moment at least, LEDs have the potential to give much better color rendition
than do fluorescent lamps.

## INSIDE AN LED

This book is about designing lighting with LEDs, not about how to make them.
Nonetheless, some aspects of their construction are worth knowing. It helps to
understand some of the design aspects of different manufacturers' products. It also

helps to understand some of their claimed improvements in lifetime. We'll be talking about white LEDs made with phosphors, although much of the information is the same for other types.

The first thing to realize is that while almost all of the devices currently used by engineers—diodes, transistors, logic gates, microprocessors—are made of silicon, LEDs are not made of silicon. (There used to be some germanium devices around, but they don't work very well when they get hot, and so were abandoned.) However, it has proven difficult to get silicon to emit light. Thus, a number of different semi-conductors have been put to use. While it's not important to know the details, you should realize that there are a variety of different materials being tried. Not all of the physics is understood yet, and the aging processes are unclear as well. Different types are in use for different devices from different manufacturers. What this means practically is that you should expect changes ahead. The device you buy today will probably be different from what is available tomorrow.

The fundamental semiconductor device in an LED is relatively large, a few square millimeters. This device emits blue light (for white LEDs), and two things must be done to it: the blue light has to be converted to white light with high efficiency, and the white light has to come out without being blocked. So the normal ceramic package that ICs come in won't work, because it (intentionally) doesn't let any light through.

What most manufacturers do is to add some transparent silicone (a rubbery polymer) on top of the die. This lets the light come out without much absorption or color change, bending the light as needed, and providing a degree of mechanical protection for the die. At least one manufacturer then adds a piece of glass on top of the silicone, although it's not clear to us that this offers much advantage.

To accomplish the color conversion, a phosphor is used, which is a complex molecule that absorbs the blue light that the LED is emitting and radiates it out over a band of other colors. It takes two or three different phosphors to make a reasonable white color; you should expect to see phosphor blends with even more components in the future.

Some manufacturers put the phosphor directly on top of the die, with the silicone going on top of that. Others stir it into the silicone before putting the mixture on top of the die. Putting it directly on the die increases the amount of blue light that is absorbed, but makes the phosphors sit at the same temperature as the die. Phosphors tend to degrade with high temperature. Indeed, phosphor degradation is one of the major reasons why LED light output decreases with age. Putting the phosphor in the silicone reduces the temperature the phosphors have to survive, but decreases the amount of blue light that is absorbed and converted. You could add more phosphor to compensate for this, except that phosphors are relatively expensive.

The die, phosphor, and silicone are all in a package. (And every manufacturer has its own package and footprint.) The package includes bond wires that connect the die to the leads so that you can put current through the LED. Even though it's just a single device, multiple bond wires are used in parallel to accommodate the relatively high currents (Fig. 1.3).

Now the package has an unwanted side effect. Since the LED emits light over a broad angle, some of the light is intercepted by the package. This affects efficacy

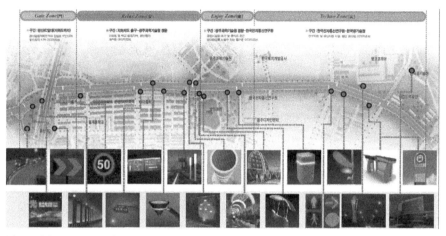

Figure 1.3   LEDs can be used everywhere. (*Source:* Kaist, KAPID.)

somewhat, but also some of the intercepted light is reflected and emitted. That's okay, except that as the package ages (it's sitting at 85 °C for 50,000 h), it yellows. As the package yellows, the absorption of light by the package increases, which decreases the efficacy. And the reflected light is also yellowed, causing the correlated color temperature (CCT) and CRI of the emitted light to shift. In some devices, this package aging is one of the major reasons why the LED time to 70% light output is 50,000 h and not longer.

Some LEDs also include some optics in their package. This may take the form of a lens and/or a mirror. The optics may be used to increase light extraction or to shape the emission direction of the light. If you don't care about the emission direction of the light (e.g., if you're building an omnidirectional light bulb), you should try to avoid using devices with extra optics. (Why pay for the extra cost?)

Thus, LEDs are complicated devices. It's well worth your while to ask detailed questions of your vendor about how the devices are made and how they will stand up to high-temperature aging. You may even need to speak to people at the factory to get sufficient information.

## IS AN LED RIGHT FOR MY APPLICATION?

To listen to enthusiastic marketing, it seems that LEDs can be used everywhere. But even though this book is about LEDs, we have to acknowledge that not every application will be best served by them. As LEDs continue to increase in efficacy and drop in price, more and more applications will benefit from them. We expect that ultimately fluorescent tubes will become obsolete. But we also expect that incandescent bulbs will be around for a long, long time. Here's a checklist of things to think about in deciding whether an LED solution is right for your application (Table 1.1).

**TABLE 1.1   Checklist of Considerations on Whether to Use LEDs for an Application**

| Question | LED | Fluorescent | Incandescent |
|---|---|---|---|
| Is energy efficiency top priority? | LEDs are probably best | | |
| Is cost an important factor? | | Fluorescents should be considered | |
| Is cost the *only* thing that matters? | | | Best to use an incandescent |
| Does the application need long life? | LEDs, properly designed, are the best choice | Fluorescents may be good enough | |
| Are there lots of on/off cycles? | LEDs should definitely be used | | |
| Are there temperature extremes? | LEDs are better than fluorescents, and usually good enough | | For really extreme conditions, incandescent bulbs are even better |
| Is the heat generated used for other purposes? | LEDs may not dissipate enough heat, for example, to melt snow off a traffic light | Fluorescents also may not dissipate enough heat, for example, to melt snow off a traffic light | Incandescent bulbs may remain a good choice |
| Is good color rendition needed? | LEDs are sometimes good enough | Fluorescents almost never are | Incandescent bulbs remain the best |
| Do colors need to be changed in operation? | LEDs are the only choice | | |
| Is a new form factor needed? | LEDs are the only choice | | |

# HAITZ'S LAW(S)

You've probably heard of Moore's law. This was the prediction by Moore in 1965 that the performance of microprocessors would double every 2 years. It was based on observations, but proved to be remarkably accurate for the next 40 years. It is only now that it has finally slowed, as ICs reach some fundamental physical limits.

A similar prediction for LEDs was made by Roland Haitz (2006). This is backed by much more historical data (see Fig. 1.4). As currently stated, it predicts that the luminous output of individual LED devices is increasing at a compound rate of 35% per year and that the cost per lumen is decreasing at 20% per year. To the extent that current manufacturers seem to have settled on 3 W as the maximum practical power in a small device, we can read this as also meaning an increase in efficacy of 35% per year.

This predicted rate of performance increase would be utterly unbelievable, except that it appears to be true. The authors began tracking the prediction a number of years ago, calculating where efficacy would be each month. Year after year, we have verified the numbers, and efficacy indeed continues to increase.

We talked to Haitz a couple of years ago about his law. His opinion was that it still had a long run ahead of it. And while he may be right that the lumens per device will continue to increase, in the next few years the efficacy will certainly start deviating, of course due to fundamental physical limitations.

To understand Haitz' law, we need to consider the meaning of "lumens" (see Chapter 3 for more detailed information). Lumens is not exactly a measure of light, but is rather a measure of how much light humans see with their eyes. As such, it very much depends on how eyes work. In particular, human eyes are most sensitive to green light. Thus, if you produce 1 W of light at 555 nm, you have 683 lumens. There's no possible way to increase this number; it is really almost a definition. The same is of course true for LEDs. If an LED gets 1 W of power, and converts it entirely

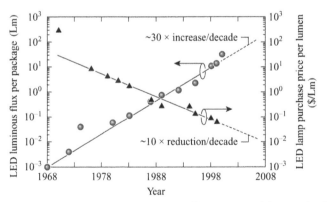

Figure 1.4  Haitz's law. (*Source:* http://i.cmpnet.com/planetanalog/2007/07/C0206-Figure3 .gif. Reprinted with permission from Planet Analog/EE Times, copyright United Business Media, all rights reserved.)

to light at 555 nm, it will have no heat power dissipation at all. (Obviously, this is not really possible because of the Second Law of Thermodynamics.) All of that light then is equal to 683 lumens. So efficacy is limited to 683 Lm/W no matter what.

Now the reality is that we don't normally want intense green light. We want white light. And since white light consists of many different colors, the lumens and efficacy must be less than 683 Lm/W. What then is the real limit on efficacy?

There are two different limits, depending on how the white light is generated. Recall that white light can be made either by directly combining lights of different colors or by emitting low-wavelength light (such as blue or UV) and converting it with phosphors into white. The phosphor method is limited by the physical efficiency of phosphors. Since they absorb low-wavelength (=high energy) light, and emit higher wavelength (=lower energy) light, the difference in energy is lost as heat. This is described by Stokes' law. While the exact limit is subject to details (such as what CRI light is acceptable), phosphor conversion of white light from LEDs is limited to about 238 Lm/W. Note that since fluorescent tubes are also phosphor-converted, but starting from 235 nm rather than 435 nm, their ultimate efficacy is considerably lower than that of LEDs. While they too have room for improvement currently, ultimately LEDs will be more efficient than fluorescent lighting.

Direct emission of various color lights can be more efficient, because there is no absorption and re-emission involved. But since colors other than green are needed, the human eye response means that 683 Lm/W isn't achievable this way either. A seminal paper by Ohno (2004) shows that to get acceptable CRI, white light cannot be made at higher efficacy than about 350 Lm/W.

Haitz's law as extrapolated to efficacy thus has several more years to run. As the 200 Lm/W limit is reached, blue converted by phosphors will plateau in efficacy. To continue increasing in efficacy, red-green-blue (RGB) systems will need to be implemented. But if 35% per year is continued, only 2 years will remain before the ultimate limit in efficacy is achieved. After that, Haitz's law may still apply to the cost per lumen. Indeed, the figure shows that it is not until after 2015 that the cost per lumen of LEDs will approach that of 60 W incandescent bulbs.

Ultimately, then, LEDs can be expected to reach the theoretical limit of efficacy, and their cost can ultimately drop below that of incandescent bulbs. And what happens after that? Since efficacy can't be increased because of physics, it might be reasonable to suppose that LEDs are here to stay for the long term. Nothing can be better than LEDs, only cheaper.

## THE WILD WEST

The LED lighting industry, and the LED industry in particular, is currently like the "Wild West": There aren't many rules, and most people aren't paying attention to them anyway. All sorts of claims are being made that are obviously wrong, and plenty more that you need special equipment to detect.

Looking first at LED device production, we should start out by saying that there are *some* reputable manufacturers. These tend to be the largest ones, although you can't assume that that's true either. They produce what they say they do, and their datasheet contains information from measurements they've taken. The problems start rather with their marketing departments.

The biggest players are presently in a contest to demonstrate that they have higher luminous efficacy white LEDs than their competitors do. As a result, they routinely release press announcements proclaiming their progress. Now everyone in the industry measures efficacy at a temperature of 25 °C. That's just a given. But actual operation is always at elevated temperatures, since LEDs heat up in operation. And the press announcements never mention how much that wonderful efficacy rolls off at higher temperatures. Different manufacturers' processes have different roll-offs, so you don't know what you would get from this new device. What's more, it's routine to announce results from a single lab device. It's not in production, and very possibly not producible without major changes. So it's all a bit of a cheat.

Moving on, even the big manufacturers tend to have problems with efficacy roll-off with aging, which is to say, lifetime. The truth is that the various manufacturing processes appear to create LEDs that age differently. And the aging varies greatly depending on the drive current, the die temperature, and even the package temperature. The fundamental problem is that 50,000 h is 8 continuous years. There's a new LED process a couple of times a year (recall the 34% increase in efficacy per year). So there isn't time to collect data before the part is obsolete. You would think that you could extrapolate data from, say, the first 1000 h. But the truth is that this works so poorly that the committee writing the specification for LED aging gave up on it. LED lifetime? It's anybody's guess.

Further, many of the LED manufacturers have problems with data. We've seen datasheets for products that have been in production for a year that still have forward voltage copied from a competitor's datasheet. We see efficacy numbers that came from handheld meters. In some cases, the parts don't match the datasheets either in color or in efficacy. The sad story goes on and on. Thus, "Caveat emptor!" The only way to be sure of what you get is to measure it yourself. Read Chapter 14 to find out how.

Moving on now to LED bulb manufacturers, the situation is even worse, if possible. We tested a couple dozen different bulbs. Only 5% of them generated the lumens they claimed, with a majority of them being wildly off! In some cases, it was apparent that no measurement had been made at all. They calculated that each LED is rated at 60 lumens, and they put three of them in the bulb, and so the package says it is 180 lumens! No thought had been given to the drive current, the optics, the packaging, not to mention the temperature effects. The U.S. Department of Energy is making efforts to clean this up. We hope for progress in this area.

We feel that all of these problems are characteristic of an infant industry. Doubtless. all of this will improve. We just hope that consumers aren't so disappointed early on that the industry never gets to maturity.

## LEDs AND OLEDs AND . . .?

Incandescent bulbs replaced candles and kerosene lamps. Fluorescent tubes replaced incandescent bulbs for many purposes. It seems likely that LEDs will replace both fluorescent tubes and incandescent bulbs. What's next after LEDs?

There's been a lot of talk about OLEDs being the next big thing in lighting. The "O" in the front of the acronym OLED stands for "organic." But it's really still an LED. The difference is that this particular type uses organic rather than inorganic material. The OLEDs' claim to fame is that they are more mechanically flexible than inorganic LEDs. Perhaps they could be made directly into light bulb shapes or printed onto mechanical forms of light bulb shape.

As we indicated in the section on Haitz's law, LEDs are probably going to reach the maximum theoretical limit for efficacy of any light source. So if OLEDs are going to supplant LEDs, it can't be on the basis of efficacy, because it's impossible to be better. The same is true for any other new light source. Once the theoretical limit is reached, nothing can be better.

The way that OLEDs *could* supplant LEDs is if they were cheaper. Once there are a variety of possible ways of achieving the maximum efficacy, the market will ensure that the cheapest one is the one that dominates. In our view, OLEDs are really just another type of LED, and their progress is part of Haitz's law. So we don't know if OLEDs or LEDs will prove the eventual cost winner. But our opinion is that there probably won't be any newer technologies for lighting that end up completely replacing LEDs. LEDs will end up being so inexpensive that cheaper won't matter to consumers. We think LEDs are here to stay.

CHAPTER 2

# LIGHT BULBS AND LIGHTING SYSTEMS

THIS BOOK is about lighting design with LEDs. While the rest of the book is about the LED part, in this chapter we present some background on the lighting part. The reason for this is that light bulbs have been around for more than 100 years. In that time, there have been many people working on them, and much technology has been developed. While we can't claim that this is a comprehensive survey, there's probably information in this chapter that you'll be happy you have.

A few words about terminology are in order. Wikipedia[1] says that "A lamp is a replaceable component such as an incandescent light bulb, which is designed to produce light from electricity." As you can see, there is a general confusion about what to call light-producing devices. Most consumers call the device a light bulb, and the unit that holds it a lamp. Manufacturers usually call the device a lamp, and the unit holding it a fixture. In this book, we will usually try to follow consumer usage. But the reader should be aware of the difference when reading publications.

## LIGHT SOURCES

### Incandescent

Light-emitting diodes are merely the newest in a long list of different types of lighting devices. Ignoring truly ancient devices such as candles, all of them use electricity. The first and still most common light source is incandescent. An incandescent bulb works by heating a piece of metal, the filament, until it glows. By adjusting the power level, it can be made to glow different colors. The typical incandescent filament runs at about 2850 K, resulting in the familiar yellow color.[2] When you dim the bulb it receives less

---

[1] http://en.wikipedia.org/wiki/Lamp_(electrical_component), under license. Accessed December 2010. http://creativecommons.org/licenses/by-sa/3.0/

[2] Could a consumer device that runs at half the temperature of the surface of the sun be introduced to the market today? And by the way, fluorescent tube plasma runs at 1100 K, much hotter than your oven.

---

*Practical Lighting Design with LEDs*, Second Edition. Ron Lenk and Carol Lenk.
© 2017 by The Institute of Electrical and Electronics Engineers, Inc. Published 2017 by John Wiley & Sons, Inc.

power. This not only produces less light, it also reduces the temperature of the filament. This is why dimmed incandescent bulbs look reddish.

Note that the glass shell in an incandescent bulb is used to maintain a partial vacuum, preventing the filament from oxidizing and failing. There has been some research into altering the mixture of remaining gas in the shell to enhance bulb life.

The incandescent bulb runs very hot. The surface temperature of a 40 W bulb runs about 120 °C. That's why you have to wait a bit after turning it off to touch it. The common failure mode for an incandescent is for its filament to break. This typically happens after about 1000 h of operation. Switching incandescent bulbs on and off a lot can also cause the filament to fail, but in typical operation this is not the dominant reason for failure.

Before leaving the topic of incandescent light sources, a comment on safety is in order. If you unscrew an incandescent bulb from its socket without turning off the light switch, sticking your finger in the socket will connect you with 120 VAC. This is life-threatening. If you try to hold an incandescent bulb when it's on, it will burn you. It's hard to imagine a device with these sorts of extreme problems being introduced today. Conversations with engineers at UL suggest that incandescent bulbs are "grandfathered in." They were there before regulations existed, and so they can't be easily eliminated. But it certainly seems like the time has come for engineers to come up with something better.

## Halogen

Halogen bulbs are also incandescent. The difference between halogen and normal incandescent bulbs is that halogens contain a small amount of halogen. The halogen makes the filament burn hotter, which slightly increases the efficacy of the bulb. It also makes the CCT higher than in a normal incandescent. An additional benefit is that the halogen helps the filament to survive longer (by redepositing the filament material).

## Fluorescent

Fluorescent bulbs work entirely differently from incandescent bulbs. They too have a partial vacuum inside a glass tube. In this case, though, the tube intentionally has some mercury vapor in it. When the filament inside the bulb is heated, it emits electrons. These ionize the mercury, forming a plasma arc at about 1100 K. The mercury emits UV light to go back to its normal state. The UV light hits a phosphor coated on the glass tube. This is the white coating on fluorescents. The phosphor absorbs the UV light, and emits visible light, which is the output of the bulb. The phosphor is carefully designed to produce just the color light that is desired, and is usually a mix of different phosphors.

To run this complicated device requires a special circuit called a ballast. The ballast is connected to the AC line as input, normally either 120 or 277 VAC in the United States. At its output the typical one bulb ballast has two pairs of wires. Each pair is heating one of the two bulb filaments. Additionally, current flows from one pair to the other. This latter is the current that produces the plasma arc. Figure 2.1 shows the currents in this lighting system.

Figure 2.1 Currents in a fluorescent tube.

Fluorescent tubes run much cooler than incandescent bulbs. The typical surface temperature is about 40 °C. You can easily touch them and pull them out of their fixture while they're running. For this reason, they typically have an electrical interlock system. If the tube is not present, the ballast typically is designed to turn off to avoid shocking you if you stick your finger into the socket.

Fluorescent tubes have a variety of failure modes. The most common failure mode is for the filaments to break. This typically happens after about 10,000 h, 10 times as long as incandescent bulbs. Because the metal of the filament runs so hot, the metal gradually burns off, weakening it. Additionally, every time the fluorescent is turned on, the sudden heating blows off some of the metal. This material lands on the glass, causing the end blackening seen in old tubes. Fluorescent ballasts also fail, but this is typically on the order of 10–25 years.

## Induction Lighting

Induction lighting is a type of fluorescent lighting that was designed to overcome the lifetime limits of the filaments in normal fluorescent lighting. Induction lighting doesn't use filaments. The energy is introduced into the plasma through a transformer. In this case, the ballast is the primary of the transformer, and the plasma arc is the load on the secondary. Coupling between the primary and secondary is through the air, so the ballast needs to be close to the bulb.

Induction lamps have a rated life of 100,000 h. Since there are no filaments to fail, the lifetime is determined by the vacuum seal on the bulb and by the time it takes the ballast to wear out. This sounds like a really good light, so why is it uncommon? We don't really know, but one has to wonder about the safety of being exposed to 13.6 MHz radiation from these systems.

## High-Intensity Discharge (HID) Lamps

High-intensity discharge (HID) lamps are fundamentally similar to fluorescent lamps. The major difference is that instead of generating UV light and converting it to visible light, these gas discharge lamps emit visible light directly. For example, sodium vapor lamps use sodium instead of mercury. Sodium emits yellow light, which is often seen in lights in parking lots. The term "HID" covers a variety of lamps that differ in the material used to generate the light, such as metal halide, sodium vapor, and xenon.

Because there is one less conversion step in these bulbs, HID bulbs tend to be more efficient than standard fluorescent tubes. They can be 100 Lm/W versus 60 Lm/W for fluorescent tubes. As with all lighting, there is a trade-off of light quality versus cost. The higher efficacy of these bulbs is offset by higher cost. And achieving higher CRI also reduces the lifetime of the bulbs, adding to cost.

# CHARACTERISTICS OF LIGHT SOURCES

All of these light sources have various pros and cons. That's why there are many different types of light source available. No one type has proven suitable for all applications. This section reviews some of the good and bad characteristics of these various light sources. This will help to give some perspective on the prospects for LED lighting.

## Light Quality

The first characteristic to consider for a light source is the quality of the light. We'll be getting into detail about light quality in Chapter 3. For the moment, we'll address two simple measures: correlated color temperature (CCT) and color rendering index (CRI). While imperfect in a variety of ways, these two give a broad overview of a light source. The CCT describes the color temperature—for example, yellow is hotter than red. CRI describes how well a variety of different colors are reproduced—for example, a CRI of 82 is better than 60, and a CRI of 100 is perfect.

Noon sunlight has a CCT of about 6500 K. A typical incandescent bulb has a CCT of 2850 K and a CRI of 100. This latter is the basic yellow light bulb, and pretty much the standard against which other light bulbs are compared. A variation of this is the "daylight" incandescent bulb. In this bulb, the glass is tinted with neodymium, which absorbs some of the red light from the filament. The result is a higher CCT and a lower CRI. But for some lighting applications, this less yellowish light may be preferred.

Fluorescent lights are available in a variety of CCTs and CRIs. The most common type, used in offices, has a CCT of about 4100 K and a CRI of about 82. This gives the cool white color supposedly best for working. It should be noted that the CRI is a misleading number when applied to fluorescent lights. (It's misleading for LEDs too, but that's a topic to be dealt with later in the book.) The reason is that CRI was designed for a spectrum of light that is smooth as a function of wavelength. But fluorescents' spectrum is full of spikes; it emits light at a number of very specific colors. These colors have been carefully picked to give a good CRI number, but that doesn't at all reflect the quality of the light as perceived by humans.

## Efficacy

The other big characteristic of light sources is their efficiency, or rather efficacy: How much light they produce for how much energy you put in. Incandescent lights are the bottom of the heap here. They are basically just big resistors. For example, a 60 W bulb produces 830 lumens, which is only 14 Lm/W. Higher power bulbs have slightly higher efficacy, but not by much. A "daylight" incandescent is even worse, only three-quarters of this efficacy, because one-quarter of the light is intentionally absorbed in the glass to change the color.

Fluorescents have considerably better efficacy. A 4 ft T8 tube produces 2700 lumens for 32 W, an efficacy of 84 Lm/W. But don't be misled. You can plug an incandescent directly in to 120 VAC, but the fluorescent tube requires a ballast to convert the power. With a ballast running about 89% efficient, input power is actually

36 W, so efficacy for this fluorescent system is 75 Lm/W. Compact fluorescent lamps (CFLs) are even lower, around 60 Lm/W.

## Timing

Timing covers both turn-on time and flicker. Incandescent bulbs are simple. When you apply power to them, they turn on within half a line cycle—at any rate, faster than your eye can see. And right away, they are at full brightness. Fluorescents are more complex. To ensure good lifetime of the tube, the filament is preheated before the plasma arc is created. This preheat time is typically 700 ms, which is quite noticeable, and sometimes stretches into seconds. Once the tube is on, it can then take minutes for it to come up to full brightness. This delay time is one of the major objections to compact fluorescent light bulbs (CFLs).

Some types of lamps have even longer start times.[3] Sodium streetlights can take minutes to turn on. Since they are turned on at dusk, which is only roughly defined by a photosensor, this turn-on time is not perceived to be a problem. HID lamps in general don't turn on again right after being turned off; you have to wait 10–15 min before you can turn it back on. This presents problems when there are power outages. Getting around this requires extra money, and so is part of system cost.

Flicker refers to what happens when a light turns off every time the AC line goes through 0 V. Incandescent lights of course are subject to this, but you don't notice it because the filament takes so long to cool down that the change in light isn't noticeable. (Fundamentally, the filament has a long thermal time constant. You can see this when you turn off an incandescent bulb. Some light continues to be emitted for a noticeable fraction of a second afterward.)

Fluorescent tubes, including CFLs, however, extinguish their plasma arc within about 100 µs. (Incidentally, this is why operating a fluorescent tube above about 10 kHz gives a 10% efficacy advantage over a 60 Hz operation.) They thus can be seen to turn on and off 60 (or 50) times per second. This produces an annoying flicker in the light. The same problem potentially affects LED lights, since they turn off even faster than fluorescent plasmas do.

The problem here is conflicting engineering requirements. Of course, you could keep the lamp on by supplying some internal power storage, such as a capacitor. But putting a capacitor on the AC line results in a bad power factor. At least for LED lamps, the U.S. government is requiring a good power factor, and so this option is not available, at least not cheaply. You could also add a capacitor at the output. Some fluorescent ballasts do include a capacitor, but since it adds cost it is uncommon.

## Dimming

Many lights are on dimmers. The way most dimmers work is simple. They disconnect the AC line from the load during part of the line cycle. This produces less power and therefore less light, albeit nonlinearly.

---

[3] http://ecmweb.com/lighting/electric_voltage_variations_arc/. Used with permission of Power Electronics Technology, a Penton Media publication.

All light sources have some types of problems with dimming. Incandescent bulbs drop in CCT as they are dimmed, making them look progressively redder. Fluorescent tubes and CFLs generically just turn off if put on a dimmer, since they perceive the missing voltage as a decrease in average line voltage. Further, reducing the line voltage applied to a standard fluorescent ballast means that not only the arc current but also the filament power is reduced. This can enormously shorten tube life. In extreme cases, fluorescent ballasts put on dimmers have been known to catch on fire.

LEDs potentially have the same problems. If they are designed to produce constant light output, they won't respond at all to a dimmer. As less of the line voltage is present, the current drawn during those times will go higher. At some point, the current being drawn during the remaining part of the line cycle will be so high that the driver has to shut down to protect itself. LED lamps could also be designed to dim as the dimmer cuts out part of the line cycle, but then they need some way of knowing what percentage of time the line is missing. Furthermore, the driver circuit needs to stay on during the time when the line is zero, and so needs some hold-up capacitance. This is potentially another power factor problem.

A more fundamental problem for both CFLs and LED bulbs is their energy saving characteristics. Most dimmers have some minimum load specified, such as 30 W. Since most incandescent bulbs are at least 40 W (in the United States), they work fine with dimmers. But CFLs and LED bulbs may be considerably less than 30 W. The dimmers don't work properly with these very light loads. They may turn on and off erratically, causing dimming not to work properly. In extreme cases, the dimmer can burn up. A solution that has been tried is to add in some extra load to maintain the dimmer power—but this then defeats the goal of energy savings. We'll talk more about dimming in Chapter 8.

## Aging

If you replace only one of multiple incandescent bulbs in a fixture, it is immediately apparent that the older incandescent bulbs have grown dimmer over time. The same is true for fluorescents and LEDs. The differences between them are in how long this aging takes, and what happens at end of life of the bulb.

When the manufacturer of an incandescent bulb states its lifetime to be 1000 h, this is the average time until a bulb stops working. As is turns out, this is also approximately the time until the bulb is about 70% as bright as when it started out. In the lighting industry, 70% as bright is considered to be the point at which a bulb is noticeably less bright. So this is good design; the incandescent bulb fails about the time that it starts getting noticeably dim.

Fluorescents are more complicated than incandescent bulbs, and thus their lifetime is more complex as well. Since fluorescent bulbs' lifetime depends on both operating hours and on/off cycles, their average lifetime depends on their usage pattern. So it may be that the recent reduction in the claimed lifetime of fluorescent tubes (from 10,000 to 8000 h) wasn't related to design changes for the purpose of reducing cost, but rather to a re-evaluation of the typical usage pattern. If you put a fluorescent tube on a motion detector circuit, its lifetime may be reduced compared with that of an incandescent due to the number of times it has to turn on.

This variation in usage produces more of a spread in lifetime for fluorescents than for incandescent bulbs. But it's even more pronounced for LEDs. Since they are semiconductors, LEDs have extremely long lifetimes, probably hundreds of thousands of hours. But they dim long before this happens. A good part run in a good design might have 50,000 h till it reaches 70% brightness. But unlike the incandescent bulb, the LED bulb keeps right on running after it reaches 70% brightness. So the U.S. government has decided that an LED bulb is "dead" after it reaches 70% brightness. That's the rating that is required on the box. But the reality for consumers is that it's going to keep on working almost forever. As a caution, we might note that no one yet has any real idea what the distribution of lifetimes for LED bulbs is going to be; not enough time has passed yet.

## TYPES OF BULBS

### Bulb Shapes

There are a surprising number of different bulb shapes available, as shown in Figure 2.2. The most common type in the world is the A shape, the standard pear shape. And in the A

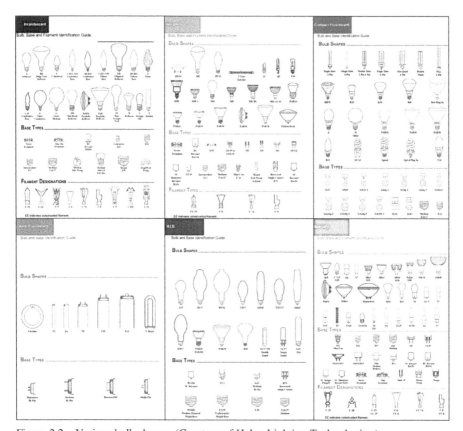

Figure 2.2    Various bulb shapes. (Courtesy of Halco Lighting Technologies.)

shape, the A19 is by far the most common. The number 19 after A signifies the diameter of the bulb in eighths of an inch (for the United States). Thus, an A19 has a diameter of 19/8 in. $= 2\frac{3}{8}$ in. In the rest of the world, the number is the diameter in millimeters, so that a common size is A55. The figure doesn't show it, but many of the bulb shapes come in various sizes. The A shape is available in A15, A19, and A21. The larger size is typically used for the higher wattage bulbs.

Other common shapes for incandescent bulbs include the BR and the PAR, which are used for floodlights and spotlights. As indicated by the name, spotlights have a narrow beam and are used to light a specific area. Floodlights have a broader beam. The G bulb is a sphere, and is commonly seen in residential bathrooms. Candelabra lights come in a variety of styles.

Halogen bulbs have a variety of shapes, some of which resemble those of incandescent bulbs. Popular types include the BR and PAR, used as substitutes for incandescent of the same shape, and MRs, which are used as spotlights. They are also used for track lighting. Track lighting comes in two types, one that uses 120 VAC directly, and the other with a ballast that converts 120 VAC to 12 V. The bulbs for this type then are designed to work on 12 V rather than 120 V.

Fluorescents come in tubes and in compact format. The most common is still the T12, although in recent years these are being replaced by T8s. The number after the "T" again refers to the diameter of the tube in eighths of an inch. A T12 has a diameter of 12/8 in. $= 1.5$ in. The normal length of a fluorescent tube is 4 ft, but there are also 8 ft types used in very high ceilings, such as the high output (HO) and very high output (VHO) tubes. There are also circlines and U-tubes, seen in ceiling panels that are too small for the normal 4 ft length of tube. CFLs are now most commonly seen in the spiral shape, but there are many multiple tube types available. (These are better referred to as "biax" and "triax.") CFLs are also being put inside a variety of incandescent bulb shapes, sometimes with their own letter designators.

## Bulb Bases

In addition to the various types of bulbs, there are also a variety of base types, but most bulbs come in only one or two base types. The most popular bulb shape is the A19 (or A55). In the United States and the European Union, this comes in a bulb base that is called the medium screw base, the E26 (or E27 in the EU). This is thus probably the most common bulb base. However, in the United Kingdom and Ireland, the A55 bulb comes in a bi-pin base. And of course, fluorescent tubes also have two pins at each end.

In California, a pin base is used for government policy reasons. Under Title 24, new and remodeled houses are required to use energy-efficient lighting. Energy efficient in practice means CFLs. But to prevent people from swapping out the CFL for an incandescent bulb, the fixtures are required to be a special type needing a pin base rather than a medium screw base. Anecdote suggests the requirement is well founded. We have repeatedly heard of people swapping out the entire light fixture after the inspector has finished inspection, to get rid of the hated CFL.

## Specialty Bulbs

There are a huge variety of specialty bulbs. We call them specialty because they are not sold in billions, but they may still be sold in tens of millions. LEDs fall in this category for the moment. For example, cars use a variety of different lights: headlights, taillights, panel lights (for the dashboard), dome lights (for overhead lighting). All of these types have been incandescent in the recent past, and presumably will be LED in the near future. Taillights on trucks and buses now are almost all LED based. Buses and airplanes use a small fluorescent tube for interior lighting. And at least in the United States, 80% of all traffic lights have converted from incandescent to LED. There's actually been a problem reported with LED traffic lights. They generate so much less heat than the old incandescent bulbs that they don't melt the snow off in the winter. We've also heard complaints from people in the Northwest Territories of Canada that without incandescent lighting, the house was not as warm in the winter.

Emergency exit signs are now nearly 100% LEDs. Here the reason is purely economic. An emergency sign is required to run for 90 min in the event of a power outage. This determines the size of the battery, the most expensive component. So in this case a change to a light source that uses less energy saves money for the manufacturer.

Similar energy-saving considerations are behind a change from incandescent lighting in refrigerator cases to LEDs. The heat generated by the incandescent is not only wasted energy; it also has to be removed by the cooling unit! There is thus a double hit on the cost.

The desire for energy saving in computers and cell phones has also led to the use of LEDs for backlighting, even though at first it wasn't entirely economical. The reason in both cases was the desire to save the battery energy. For the same reason, flashlights now almost universally use LEDs. Here the reason is not only to save the battery, but also to increase the light output without overheating. Televisions are also headed in the same direction, although for general energy-saving reasons.

Almost everything mentioned in this section has been about conversion to LEDs. One area that may not change is oven lights. Since ovens run up to 550 °F (300 °C), incandescent bulbs seem like a natural choice. They are not affected by heat, but semiconductors are. They may never be changed; there are still vacuum tubes used in some specialty areas.

# HISTORY OF LIGHTING

Edison made the first practical incandescent light bulb in the 1880s. Fluorescent tubes became popular in the 1950s. LEDs are just starting to become popular in the 2000s. What will the future bring?

Jeff Tsao, a Principal Member of the Technical Staff at Sandia National Labs, researched this question, which deserves to be more widely known (Tsao, 2010). He looked at worldwide consumption of artificial light over the last several hundred years, covering candles, kerosene lamps, gas lighting, incandescent bulbs and fluorescent tubes. Obviously, much of the older data can't be that accurate. But when plotted on a log scale, these inaccuracies aren't that significant. What was clear is that over the entire time period, the world has spent an approximately constant percentage of its GDP on lighting, 0.72%.

Assuming this continues into the future, this has a surprising consequence. One of these is that as lighting becomes more efficient, total light used will increase. The introduction of LED lighting at, say, twice the efficacy of fluorescent lighting will mean that twice as much light is used, not that half the energy is saved. Energy is saved according to the model only if GDP decreases or the cost of electricity decreases.

This conclusion seems to fit with what is actually happening in the world today. Governments reasoned about energy-efficient lighting as follows: "If you replace all of the 60 W incandescent bulbs with 15 W CFLs, the energy saved will be 45 W times the number of bulbs." But these energy savings have not been entirely realized. Anecdotally, people report that they want brighter lights, not lower energy consumption. The reasoning apparently goes like this: "My closet was lit by a 40 W incandescent. With a CFL I get the same light for only 10 W. But my closet was always dim. I can upgrade to a 15 W CFL and still save energy!"

Our expectation is this. As Haitz's law continues, the exponential increase in efficacy and decrease in cost will drive almost all lighting to convert to LED. This will be translated into a huge increase in the total amount of light used in the world. As the ultimate physical limits on efficacy are reached, the amount of light used in the world will stabilize. Increases will thereafter be small, driven by ordinary increases in GDP and population.

## GOVERNMENTS

Governments worldwide have become interested in regulating lighting. A simple calculation shows why. Twenty billion light bulbs times 60 W each times 4 h a day equals over 1 *trillion* kWh a year! This is about 20% of all electricity used in the world. This huge energy consumption has serious consequences for the economy, for national security, and for the environment.

Such considerations have led to numerous government regulations. We've already mentioned Title 24 in California. Even more draconian measures are being implemented worldwide. Many governments have banned, or shortly will ban, the sale of incandescent bulbs entirely.

Why not just let the economics speak for itself, and let people switch to energy-efficient lighting to save money? The problem is that individually the change doesn't make a lot of sense. Consider changing a 60 W incandescent bulb to a 15 W CFL (or a 10 W LED bulb). That saves 45 W. The bulb is on an average of 4 h a day. That's 180 Wh a day. With electricity costing 12 ¢/kWh, that switch to energy-efficient lighting has saved 2 ¢! While this adds up to $7.88 for the whole year, in the developed countries such an amount may not be considered worth the time spent—and many people feel it is not worthwhile going to a bad-looking light source for this amount of savings.

# LIGHTING SYSTEMS

So far we've discussed light bulbs and their usage. But bulbs aren't used by themselves; they need infrastructure to work, such as electricity and ballasts. This comes under the heading of lighting systems. Details about the electricity supply will be discussed in Chapter 8. As for ballasts, corresponding to the large number of different bulbs, there are a large number of ways to run them. In this section, we'll consider ballasts for fluorescents and then for LEDs.

As we've indicated, fluorescent tubes have special voltage and current requirements, which means they can't be powered directly by the AC line. Instead, the AC line powers a special electronic circuit called a ballast, which in turn powers the fluorescent lamp. Fluorescent ballasts come in two varieties, instant start and rapid start. The difference lies in how the filaments are treated. You'll recall that a filament is used to provide electrons to the plasma arc. In a rapid start ballast, the filaments are preheated, that is, they have power applied to them and are warmed up to about 1100 K before they have to start emitting electrons. This has the benefit of making the filaments last a long time, but also has the downside of requiring extra power. With an instant start ballast, the filaments are not preheated. This saves power, but makes filament life, and thus tube life, much shorter.

In principle, you could get the best of both worlds by preheating the filaments, and then turning them off after arc ignition. But this isn't common, probably because of the cost of the extra circuitry to do this. Something similar is done, however, with dimming ballasts. (Normal fluorescent ballasts don't work with dimmers; you have to buy a special type that does.) As the tube is dimmed, the plasma arc heating the filament heats it less and less. To maintain filament temperature, which is what keeps them running for a long life, filament power is increased in a dimming ballast inversely proportionally to the degree of dimming.

One other differentiation among fluorescent ballasts is that some are magnetic, while others are electronic. Magnetic ballasts are basically a large piece of iron in the form of a transformer, with a big capacitor on the output to limit current into the fluorescent tube. They of course run at line frequency, which is why they are big and heavy. Electronic ballasts are switch-mode power supplies. They are much lighter, but have the same mechanical size as magnetic ballasts for backward compatibility. They tend to run above 20 kHz, for reasons of size and to avoid audio noise, as well as to take advantage of the small gain in efficacy of fluorescent tubes above 10 kHz.

Most LEDs are similar to fluorescent tubes in not being directly compatible with the AC line. (Some are, and this may well be the future of LED systems.) They also require a ballast to convert AC power to the constant current that is best for LEDs. Chapter 8 covers design of these ballasts. It is to be noted that some people reserve the word "ballast" for fluorescent tube electronics, and prefer the word "driver" for LEDs.

We have the AC power line, the ballast, the lamp; surely that's everything in the lighting system? No, there's also the physical environment the lamp operates in. It too influences light production and efficacy. Since incandescent bulbs operate at thousands of degrees, they are relatively insensitive to environmental temperature; not so fluorescent and LED bulbs. Fluorescent systems require a longer time to start when it's cold. The fluorescent tube in your garage has trouble starting in winter for this reason. There are special fluorescent ballasts designed for cold weather. Conversely, CFLs can have problems with high ambient temperature. Placed into a ceiling can, the temperature can rise enough to cause the CFL to fail, sometimes catastrophically. Since the actual temperature in the can depends on many factors (how much power is being dissipated, whether the can is insulated or enclosed, what the ambient temperature is), in general CFLs are not recommended for can usage.

LEDs lights have similar problems. Although they are not as sensitive to cold, their ballasts may be. And if they get hot, their light output decreases and their lifetime also goes down. In an insulated enclosed can, they can get hot enough to fail. This can be particularly true if you design an LED system to have constant light output. As the temperature increases, their efficacy decreases. Thus, you put more power into the LEDs to compensate for their decrease in efficacy. But this causes the temperature to rise even more.

Another environmental factor is the fixture into which the bulb is placed. This can dramatically influence the light output of the lighting system. As an extreme example, consider a fluorescent tube mounted into a black fixture. Since half of the light from the tube goes upward, this half will be absorbed. The actual light output from the system will be only half that generated by the tube, thus also reducing the effective efficacy by half. While tube fixtures are not usually black, they can lose as much as 30% of the light output. LEDs can potentially avoid this loss because they are not omnidirectional.

One more environmental factor almost never discussed is dust. There is some residual voltage present on fluorescent tubes that tends to attract dust. (The same is more noticeable on your laptop computer screen.) Of course, no one dusts their lamps. Over years of operation, this dust can cut the actual light output from the lighting system by 30%. Whether LED bulbs also suffer this problem, or can claim additional efficacy over fluorescent lighting systems, remains to be seen.

CHAPTER *3*

# PRACTICAL INTRODUCTION TO LIGHT

Lɪɢнт ɪs simply electromagnetic waves whose wavelength is in the visible range. The energy associated with light waves is measured in watts. The quantities used to characterize light physically are *radiometric*. The way light looks, for example, what color or how bright is it, depends on the human eye and brain. The human eye responds to roughly 380–780 nm wavelengths. But it gets quite complicated since the eye's response varies depending on relative conditions and differs from individual to individual. The quantities used to characterize how light is perceived by humans are *photometric*.

A light source is characterized by its emission spectrum. Lighting engineers care about the amount of light and the color of the light, not just from the emitter but also from absorbing and reflecting surfaces illuminated by the light. Most importantly, we must understand how the end user perceives the light.

To architects, often only lumens and color temperature are specified. To cockpit instrument designers, light intensity is the most important factor. To LCD screen engineers, brightness is the most important measure. To lighting designers, irradiance is the important quantity. But to the consumer, color rendition may be of greatest importance. We shall cover all these concepts at a practical level for the LED lighting engineer.

## THE POWER OF LIGHT

### Background: Light as Radiation

The energy produced by light (e.g., of a certain wavelength) is measured in joules; the rate of joule usage is given in watts. This is the radiometric quantity, the radiated energy per second. The energy is proportional to the frequency of light via the Planck–Einstein equation, $E = hf$, where $h$ is Planck's constant and $f$ is frequency (Fig. 3.1).

Most of the time, light is referred to by its wavelength. For example 475 nm is blue light and 650 nm is red light. Since frequency $f$, wavelength $\lambda$, and speed of

*Practical Lighting Design with LEDs*, Second Edition. Ron Lenk and Carol Lenk.
© 2017 by The Institute of Electrical and Electronics Engineers, Inc. Published 2017 by John Wiley & Sons, Inc.

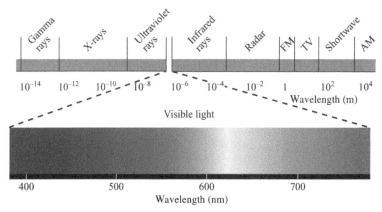

Figure 3.1   The electromagnetic spectrum.

light $c$ are related by the equation $f\lambda = c$, we have $E = hc/\lambda$. Therefore, the shorter wavelength blue light has higher energy than the longer wavelength red light. If we extend the spectrum farther out, even shorter wavelength than blue light is ultraviolet light at 230–400 nm. UV has much higher energy than visible light. On the other end of the spectrum is infrared light, with a wavelength longer than that of red light, at 700 nm–1 μm.

Light of a single frequency is *monochromatic* light, such as that from a laser or a color filter. But most sources of light have a mix of colors. The radiant power of such a light is the sum of all of the power at each frequency.

When referring to light sources, the term "radiant power" is often replaced by the term *radiant flux*. Flux refers to total power radiating out in all directions, regardless of how much area it intercepts. Imagine a candle inside a balloon. The inside surface of that balloon sees all the flux of the candle. If the balloon is blown larger, it would still see the same flux.

## RADIOMETRIC VERSUS PHOTOMETRIC

So far we've only talked about the light source, which is measured by radiant flux. Since we are only concerned with how that light is perceived by the human eye, we use the radiant flux as weighted by the spectral response of the eye to derive the *luminous flux*, measured in lumens (Lm). All quantities weighted by the eye response are photometric quantities. For every radiometric measure, there is an equivalent photometric measure. Photometric quantities are measured in "lumens." The word "photometric" refers to the visible range of light to which the human eye is sensitive.

Figure 3.2 shows the eye response curve as standardized by the Commission Internationale de l'Eclairage (CIE). Photometric quantities are weighed by *photopic vision*, or bright light vision, as opposed to night vision, or *scotopic vision*. You'll notice that the photopic vision peaks at 555 nm yellow green. Human eyes are most sensitive to this color green. This is why a green laser pointer looks much brighter compared with blue or red laser pointers of the same power (Table 3.1).

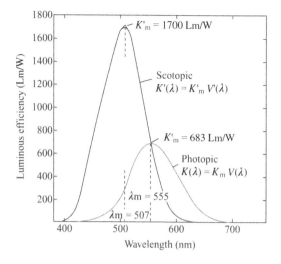

Figure 3.2   Scotopic vision is much more sensitive than photopic vision. (*Source*: Kalloniatis and Luu (2007).)

**TABLE 3.1   Radiometric versus Photometric Description of Light Power**

| | Quantity | Description | Unit | Symbol |
|---|---|---|---|---|
| Radiometric | Real power | Radiant flux, radiant power | Watts (W) | |
| Photometric | As seen by humans | Luminous flux | Lumen (Lm) | $\Phi$ |

Scotopic vision has a similar shape sensitivity curve, but is shifted toward the blue, shorter wavelength side by 45 nm, to peak at 510 nm, or pure green. Scotopic vision is also much more sensitive to light than photopic vision. Its peak efficacy is around 1700 Lm/W, as opposed to 683 Lm/W for photopic vision. A 15 Lm light won't look as bright as it looks in a dark room where the scotopic vision takes over.

Intermediate between scotopic and photopic vision is a level of light where both are active simultaneously. This is called *mesoscopic* vision. An example of mesoscopic light levels is the light outdoors at night, where there are a lot of background sources at a distance. If you are designing for mesoscopic lighting conditions, remember that the usual lighting levels are specified for photopic vision. This may result in producing more light than is actually necessary for the mesoscopic application.

The conversion between radiant and luminous flux depends on the wavelength spectrum of the light. Returning to photopic vision, if we take a 1 W green light at the peak of the human eye response, 555 nm, then the luminous flux is 683 Lm. At this peak, there is 100% conversion from radiated optical watts to lumens. This is the maximum efficacy in the sense that 1 W of any other wavelength or color of light will convert to fewer lumens.

Let's take a look at how the spectrum affects the conversion from radiant flux to luminous flux. Figure 3.3 shows the spectra of four light sources. Incandescent light bulbs are black body radiators and weigh heavily toward the long wavelength (red) end and into the infrared range. Compact fluorescent light bulbs have a spike at the

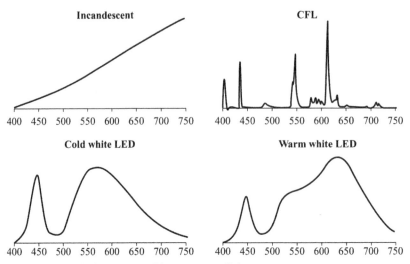

Figure 3.3  Emission spectra of four common light sources.

edge with ultraviolet and then spikes at each of the phosphor wavelengths. The warm white LED has a phosphor emission that is more toward the red end when compared to the cold white LED.

Let's compare radiated power of each light to the number of lumens. Table 3.2 lists some typical values. In the last column, we see that the incandescent bulb has the lowest conversion factor from radiant flux to luminous flux. This is as we expected, since its spectrum was the worst match for the eye response spectrum. Another interesting thing to notice is that CFLs and LEDs are twice as efficient as incandescent bulbs, but more than four times as efficacious due to the closer matching between their spectrum and the eye response. Obviously, CFLs have a narrow spectrum and consist of only a few frequencies. LEDs, on the other hand, are much wider spectrum. This difference has implication in the color rendition, as we shall see later in this chapter.

**TABLE 3.2  Radiant and Luminous Flux**

| Light source | Input power | Color temperature | Radiant flux | Efficiency | Luminous flux | Efficacy | Luminous flux/radiant flux |
|---|---|---|---|---|---|---|---|
| Symbol | $P$ | CCT | $\Phi_{rad}$ | $\Phi_{rad}/P$ | $\Phi_{lum}$ | $\Phi_{lum}/P$ | $\Phi_{lum}/\Phi_{rad}$ |
| Units | W | K | W | | Lm | Lm/W | Lm/W |
| Incandescent | 58 | 2700 | 5.38 | 9.3% | 749 | 12.9 | 139 |
| CFL | 9 | 2600 | 1.50 | 16.6% | 511 | 56.8 | 342 |
| CW LED | 7.5 | 3900 | 1.77 | 23.7% | 619 | 82.6 | 349 |
| WW LED | 8 | 3050 | 1.83 | 22.6% | 518 | 64.1 | 284 |
| 555 nm Green light | — | — | 1 | — | 683 | — | 683 |

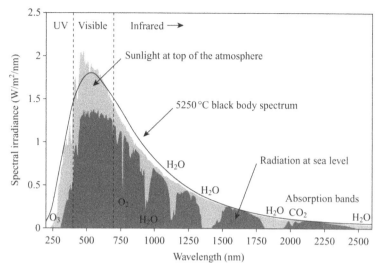

Figure 3.4  Solar radiation spectrum. (*Source:* http://en.wikipedia.org/wiki/File:Solar_Spectrum.png under license http://creativecommons.org/licenses/by-sa/3.0/. Accessed January 2011.)

**TABLE 3.3   Incandescent Efficacy Increases with Wattage**

| Wattage | Initial lumens | Initial efficacy (Lm/W) |
|---------|----------------|-------------------------|
| 40 | 505 | 12.6 |
| 60 | 865 | 14.4 |
| 75 | 1190 | 15.9 |
| 100 | 1710 | 17.1 |

For comparison, the solar radiation spectrum is shown in Figure 3.4. It is fairly evenly distributed save for the violet colors near the UV.

An incandescent bulb radiates light approximately evenly in all directions (it is *isotropic*). This qualifies it as a perfect light source to be specified in lumens. In fact, some boxes of bulbs do list the lumens along with the usual wattage. In Table 3.3, you'll notice the efficacy is higher in the higher wattage incandescent bulbs. This is due to the higher temperatures of the filament, and thus the higher black body radiation efficiency.

## LUMINOUS INTENSITY, ILLUMINANCE, AND LUMINANCE (OR CANDELA, LUX, AND NITS)

Unless you're an expert in lighting, chances are very good that you do not know the difference between luminous flux, luminous intensity, illuminance, and luminance. We're going to go into detail in this section to remove the confusion, but here is a summary. *Luminous flux* is how much light is perceived by humans as coming out of a source, and is measured in lumens. *Luminous intensity* is how much light is coming out into a given solid angle, and is measured in candela = lumen/steradian. *Illuminance* is how much light is coming out onto a given area, and is measured in Lux = lumen/m$^2$. And finally, *luminance* is how much light is coming out onto a given area in a given solid angle, and is measured in candela/m$^2$ = nits.

## Luminous Intensity

The quantity of luminous intensity ($I_{lum}$) adds a directional component to luminous flux. It is the amount of light radiated into a certain two-dimensional angle. This two-dimensional angle is called the solid angle and is measured in *steradians* (sr). The solid angle traces out a cone radiating out from the source. Using SI units, a unit sphere of 1 m radius has a surface area of $4\pi$ m$^2$. So a sphere is $4\pi$ steradians, half a sphere is $2\pi$ sr, and so forth. One steradian intersects the unit sphere in a circular bowl shape. When that bowl has a surface area of 1 m$^2$, then the solid angle is defined as 1 sr. Or more generically, a sphere of radius $r$ is defined to be intersected by a solid angle of 1 sr when the surface area described by that solid angle is $r^2$ (see Fig. 3.5).

The SI unit of luminous intensity is the *candela* (cd). One candela is equivalent to one lumen/steradian. One candela is roughly the luminous intensity of a candle. Since 1 cd is 1 Lm/sr, to find the lumens in one complete sphere, we multiply by $4\pi$. A candle thus emits roughly 12.6 Lm.

A typical 100 W incandescent bulb emits 1700 Lm at start of life. These lumens are radiated over the entire sphere of solid angle, or $4\pi$ sr. To calculate the luminous intensity for this bulb, we divide 1700 Lm by $4\pi$ sr to get 135 cd.

Let's take the same bulb and put a perfect reflector behind it such that all the light is emitted into a single hemisphere, a solid angle of 180° or $2\pi$ sr. Then the luminous intensity is 1700 Lm/$2\pi$ sr = 271 cd. It is double the intensity because the same amount of light is going into half the area.

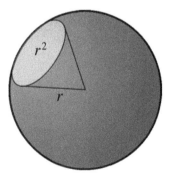

Figure 3.5   One steradian intersects 1 m$^2$ of area of a 1-m radius ball. (*Source:* http://commons.wikimedia. org/wiki/File:Steradian.png under license http:// creativecommons.org/licenses/by-sa/3.0/.)

| Light source | Luminous flux (Lm) $\Phi_{lum}$ | Angle $2\theta$ | Half angle $\theta$ | Solid angle (sr) $\Omega$ | Luminous intensity (Lm/sr = cd) $I_{lum} = \Phi_{lum}/\Omega$ |
|---|---|---|---|---|---|
| Candle | 12.6 | 360° | 180° | $4\pi$ | 1 |
| 100 W incandescent | 1700 | 360° | 180° | $4\pi$ | 135 |
| 100 W incandescent with 180° reflector | 1700 | 180° | 90° | $2\pi$ | 271 |
| 100 W incandescent with 36° reflector | 1700 | 36° | 18° | 0.3075 | 5528 |

For an arbitrary angle $\theta$, the solid angle $\Omega$ can be calculated by the formula $\Omega = 2\pi (1 - \cos(\theta))$. Be careful to define this angle as half of the total angle. For example, in the hemisphere, the beam angle $2\theta = 180°$, so that the half angle $\theta = 90°$, and thus $\Omega = 2\pi(1 - \cos(90°)) = 2\pi$ sr. At a beam angle of 60°, $\theta = 30°$, and we find $\Omega = 2\pi (1 - \cos(30°)) = 0.84$ sr. See Figure 3.6 for a graphical representation of this function.

Now let's take that same 100 W bulb and extend the reflector such that all the light is emitted into a 36° flood light. Setting $\theta = 18°$, $\Omega = 0.3075$ sr, and $I_{lum} = 5528$ cd.

Lower power indicator style LEDs that emit light mostly in a forward pattern will be specified in candelas rather than lumens. For example, an LED might have a *view angle* of 15°, intensity of 13 cd, and flux of 0.7 Lm. All of the lumens are emitted within the view angle. So with $\theta = 7.5°$, we get $\Omega = 0.05$ sr, and $I_{lum} = 0.7$ lm/0.05 sr $= 14$ cd. The 1 cd difference between our number and theirs is probably due to rounding, since there's only 1 significant figure.

Let's say that you want to build an MR-16 equivalent light bulb with LEDs. Table 3.4 shows common types of MR-16 bulbs and their lumen output. *Beam angle* is defined by the point where the intensity drops to 50% at maximum intensity (see Fig. 3.7). A 20 W, 12° spot light has 3350 CBCP (center beam candle power). "Candlepower" is an old term for candela. So let's first calculate how many lumens

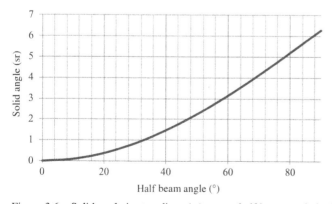

Figure 3.6   Solid angle in steradians (sr) versus half beam angle in degrees (°).

**TABLE 3.4 Common Types of MR-16**

| MR-16 | Beam angle | $I_{lum}$ (candela or CBCP) | Solid angle $\Omega(sr) = 2\pi$ (1-cos(beam angle/2)) | Luminous flux (lumens) $\Phi = I_{lum} \times \Omega$ |
|---|---|---|---|---|
| 20 W spot | 12° | 3350 | 0.0344 | 115 |
| 20 W narrow flood | 24° | 880 | 0.1373 | 121 |
| 20 W flood | 36° | 450 | 0.3075 | 138 |
| 35 W narrow flood | 24° | 1700 | 0.1373 | 233 |
| 35 W Flood | 36° | 950 | 0.3075 | 292 |
| 35 W wide flood | 60° | 370 | 0.8418 | 311 |
| 50 W spot | 12° | 8600 | 0.0344 | 296 |
| 50 W narrow flood | 24° | 2700 | 0.1373 | 371 |
| 50 W flood | 36° | 1500 | 0.3075 | 461 |
| 50 W wide flood | 60° | 620 | 0.8418 | 522 |

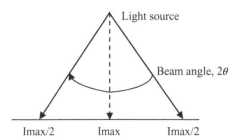

Imax/2    Imax    Imax/2          Figure 3.7   Definition of beam angle.

are emitted. We set $\theta = 6°$ to get $\Omega = 0.0344$ sr. Then we multiply $I_{lum} \times \Omega = 3350$ cd $\times 0.0344$ sr $= 115$ Lm. Remember that candela $=$ lumens/sr. So, for an equivalent system, we need to have the LEDs emit 115 Lm into a 12° beam angle.

We summarize the various measures of light so far in Table 3.5.

**TABLE 3.5 Measures of Light**

| Quantity | Symbol | Radiometric | Unit | Photometric | Unit |
|---|---|---|---|---|---|
| Amount of light | $\Phi$ | Radiant flux, Radiant Power | Watt (W) | Luminous Flux | Lumen (Lm) |
| Intensity of light | $I = \Phi/\Omega$ | Radiant intensity | Watts/steradian (W/sr) | Luminous intensity | Lumens/steradian $=$ candela (cd), or candlepower |
| Light on a surface | $E = \Phi/A$ | Irradiance | W/m$^2$ | Illuminance | Lux |

## Illuminance

So now let's take a look at how this light illuminates a surface, such as a book, a picture on the wall, or a keyboard. The amount of light $\Phi$ landing on a surface of area $A$ is defined to be *Illuminance* $E_{lum}$ and is measured in $Lm/m^2$ or *Lux*. In the English unit system, illuminance is measured in foot-candle (fc) which is $Lm/ft^2$. The conversion factor is fc $= 10.76$ Lux.

Taking our MR16 example above with 115 Lm, let's calculate the illuminance. The solid angle was 0.0344 sr, which is $0.0344/4\pi = 0.0027$ of the total solid angle of a sphere. Half a meter away, a sphere of radius 0.5 m has a surface area of $\pi\,m^2$, and so the area illuminated by the MR16 is $0.0027 \times \pi\,m^2 = 0.0086\,m^2$. The illuminance is thus $115\,lm/0.0086\,m^2 = 13,372$ Lux. You should note that a Lux meter is less than \$100 and is the easiest way to measure light. However, the meter is quite sensitive to the exact distance to the source via the inverse square law. So, if the Lux meter moves 1 cm farther away, then its reading will change to $13,372$ Lux $\times\,[(0.50)^2/(0.50 + 0.01)^2] = 12,853$ Lux.

As another example, consider the following. Our company wants to build a USB light for lighting up a computer keyboard and for reading printed material. The USB port can run an LED at 37 Lm (see Chapter 11 for this number and its usage in a real design example). Is this bright enough?

For this calculation, we need three numbers. We will take the distance from the LED source perpendicularly down to the keyboard to be $d = 0.33$ m (see Fig. 3.8). We will take the larger edge of the keyboard to be $x = 0.3$ m. The smaller edge we will take to be $y = 0.12$ m.

Now, the key to the calculation is to know how the LED light is dropping off as you get further away. In general, the *Lambertian* source (Fig. 3.9) can be adequately modeled as a cosine: At $0°$ it is 1.00, and at $90°$ it is approximately zero. Without going into all the math, the answer turns out to be that the part of the light that is intersected by the keyboard will be proportional to $x^2/(x^2 + 4\,d^2)$. We thus find that the

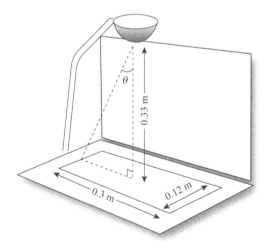

Figure 3.8   Typical Lambertian radiation pattern. (*Source*: Technical Datasheet DS56, Power Light Source Luxeon Rebel, Philips Lumileds Lighting Co., 2007.)

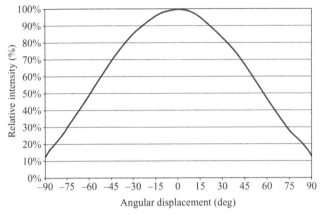

Figure 3.9   Dimensions for a USB keyboard light design.

luminous flux on the keyboard is $37 \, \text{Lm} \times (0.3 \, \text{m})^2/[(0.3 \, \text{m})^2 + 4 \times (0.33 \, \text{m})^2] = 6.34$ Lm. In other words, the size of the keyboard is small compared to the height of the LED above it, so that only 17% of the light emitted by the LED is shining on the keyboard. This is then fairly poor efficiency usage of the light. Since we know the total luminous flux on the keyboard, the illuminance is the flux divided by the area. And thus we have illuminance $= 6.34 \, \text{Lm}/(0.3 \, \text{m} \times 0.12 \, \text{m}) = 176$ Lux. Looking at Table 3.6, we see that this is enough for casual reading.

It is worth noting that if the LED were brought in closer, the efficiency would be much better. For example, if the distance were halved, $d = 0.17 \, \text{m}$ instead, and we would have $37 \, \text{Lm} \times (0.3 \, \text{m})^2/[(0.3 \, \text{m})^2 + 4 \times (0.17 \, \text{m})^2] = 16.2 \, \text{Lm}$. The keyboard then uses 44% of the light instead of 17%, more than 2½ times better.

**TABLE 3.6   Examples of Illuminance in Lux**

| Illuminance (Lux) | Example |
|---|---|
| 0.00001 | Light from Sirius, the brightest star in the sky |
| 0.0001 | Total starlight, overcast sky |
| 0.002 | Moonless clear night with airglow |
| 0.01 | Quarter moon |
| 0.27 | Full moon on a clear night |
| 1 | Full moon overhead in tropical latitudes |
| 3.4 | Dark limit of civil twilight under a clear sky |
| 50 | Family living room |
| 80 | Hallway/toilet |
| 100 | Very dark overcast day |
| 320–500 | Office lighting |
| 1000 | Overcast day; typical TV studio lighting |

## Luminance

In this book we won't actually be using luminance for any of our examples, as its usage is slightly specialized. But a brief overview of the concept is necessary to round out your knowledge of lighting. Luminance is sort of a combination of luminous intensity and illuminance. It is how much light is striking (or coming out of) a surface per unit area (like illuminance) per solid angle (like luminous intensity). Its units are thus candela/$m^2$ (also called a *nit*).

Fundamentally, luminance is a measure of how bright a surface appears to a human eye at a specified angle to the surface. Its main present-day use is in characterizing the brightness of computer monitors.

## Summary of Amount of Light

To reiterate from the opening of this section, there are four common measures of light. Luminous flux is how much light humans perceive as coming out of a source, and is measured in lumens. Luminous intensity is how much light is coming out into a given solid angle, and is measured in candela. Illuminance is how much light is coming out onto a given area, and is measured in Lux. And luminance is how much light is coming out onto a given area in a given solid angle, and is measured in cd/$m^2$.

# WHAT COLOR WHITE?

With monochromatic light sources such as lasers or nearly monochromatic ones such as color LEDs, color is a straightforward quantity. With a white LED, a continuous spectrum of different colors is mixed, as is shown in Figure 3.10. Notice how dramatically different spectra can all be considered to be white light.

The wavelength spectrum can be converted to a point on any color space to represent the resultant color as in Figure 3.11. The $(x, y)$ values are referred to as *chromaticity coordinates*. In this 1931 RGB color space, a color is represented as a

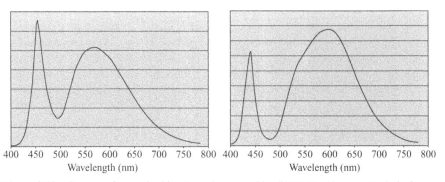

Figure 3.10 Spectra of neutral-white (a) and warm-white (b) LEDs. (*Source:* Technical Datasheet DS56, Power Light Source Luxeon Rebel, Philips Lumileds Lighting Co., 2007.)

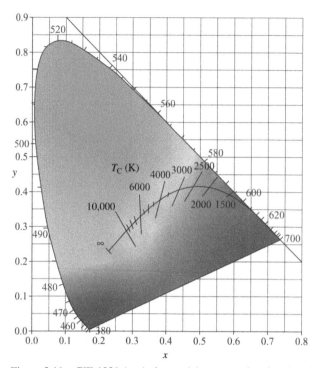

Figure 3.11    CIE 1931 (*x*, *y*) chromaticity space, showing the Planck line and lines of constant CCT. (*Source:* http://en.wikipedia.org/wiki/Color_temperature under license http://creativecommons.org/licenses/by-sa/3.0/.) (See color plate section.)

mix of varying amounts of red, green, and blue colors. The colors on this chart are highly saturated to show the colors better.

In the center of the color space, there is an area of generally white light where the red, green, and blue are added about equally. A useful reference for white light is a *black body* or *Planckian* radiator. A Planckian radiator will radiate different color light at different temperatures (think of "red hot"). The colors of these temperatures between 2700 and 20,000 K are all perceived as white light by the human eye. This forms the Planckian locus and is also plotted in Figure 3.11 as the curving line in the center. The temperatures in Kelvin are referred to as *color temperature*. For example, the sun is a black body radiator. The sun at noon on the equator has a color temperature of 6500 K. At sunset, the color temperature will be much lower and have a reddish color. Sunlight color also depends on time of day, latitude, and season.

The incandescent light bulb is also a black body radiator and has a color temperature of about 2700 K. The halogen light bulb is also an incandescent black body. The difference between them is a small amount of halogen the halogen light bulb contains, which improves its efficacy. The halogen allows it to burn hotter and therefore has a color temperature of 3000 K.

Light sources whose spectral distribution is different from a Planckian radiator will not land on the Planckian locus. They can, however, be color matched to the most similar color temperature of the Planckian locus. For these sources the term *correlated color temperature (CCT)* is used. CCT follows the *isotemperature* (equal color temperature) lines, also shown in Figure 3.11.

Fluorescent tubes for use in offices are generally 4100 K, although both lower and higher color temperature tubes exist. The term "warm white" was defined by ANSI to help consumers get a better handle on fluorescent light colors and refers to a CCT of 2700 and 3000 K. "White" is 3500 K. "Cool white" is 4100 K. "Daylight" is 5000 and 6500 K. Compact fluorescents have improved the phosphors for warm white versions so that their color temperature can be as low as 2700 K.

LED manufacturers are following the same line of thought and getting creative with terms such as "neutral white" and redefining what cool white refers to. The ANSI C78.377-2008 standard for LEDs simply lists a "nominal CCT" of 2700, 3000, 3500, 4000, 4500, 5000, 5700, and 6500 K. The names of colors are mentioned as a side comment. The lesson here is that just looking at the name of the color may not be that informative. You need to check what "CCT warm white" from a manufacturer really means.

## MacAdam Ellipses

It is important to realize that a point plotted on the color space is simply an equivalent color as perceived by the human observer. In actual experiments, a human observer is presented a source light on one side of a screen. On the other side, a light is formed by varying the amounts of red, green, and blue light until the two lights look like the same color. The limited sensitivity of the eye gives rise to an ellipsoidal area in which color differences cannot be perceived. These are called *MacAdam ellipses* (MacAdam 1942) (see Fig. 3.12) and are used as the foundation for color bins for LEDs and fluorescent lights.

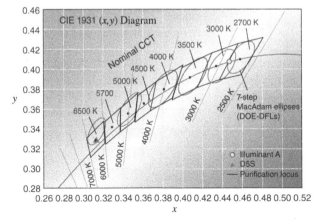

Figure 3.12   $(x, y)$ Chromaticity diagram showing CCT and seven-step MacAdam ellipses. (*Source:* http://www.photonics.com/Article.aspx?AID=34311.) (See color plate section.)

MacAdam ellipses are said to have "steps" that are standard deviations in the general human population.[1] For example, one-step ellipse means the color difference between the center of the ellipse and the edge of the ellipse can only be noticed by one standard deviation, or 68% of the population. From one edge of the ellipse to the point directly opposite is two standard deviations, or 95%. Later work, however, showed that a three-step ellipse is a more representative assessment of color discrimination. Accordingly, major manufacturers use a three-step ellipse for linear fluorescent lights.[2]

For LED color bins, ANSI C78.377, indoor lighting chromaticity chart uses seven-step MacAdam ellipses to define CCT bins. Three-step ellipses are about a quarter of the size of seven-step ellipses. LED manufacturers will sometimes split each CCT bin into further color bins. If color uniformity is important to you, ensure to choose from a single color bin.

The authors maintain that the ellipses in use today for LEDs are still too large. There are noticeable variations in color even within a single LED color bin. ASSIST (see Chapter 14) cites evidence that a four-step MacAdam ellipse is perceptually noticeable (ASSIST, 2005). This issue needs to be resolved before LED lights will be ready for widespread adoption. No one wants to buy two identical LED light bulbs and have them appear different colors. This has been an issue for CFLs. For example, some appear pinkish and some appear greenish, the difference apparently being due to small deviations above and below the Planck line. The system still isn't perfect and research is continuing. If you're building consumer lights, consider getting the tightest binning available that you can afford.

## A Note about Color Space Standards

The color space used above is based on the 1931 CIE standard. This was based on an observer with a 2° field of view. There were a variety of problems with this original research. In particular, a very small sample population was used, leading to spuriously large standard deviations (or "steps"). Furthermore, not a single observer was found to give reproducible results.

It was discovered that at a 10° field of view more of the rods were used to detect color. In 1964 a 10° field of view standard was defined. Both standards use the $(x, y)$ coordinates, so it is important to note which color space is being used. The more common standard is still that of 1931.

Another problem was also realized about the $(x,y)$ space. MacAdam ellipses in the green region are much larger than ellipses in the blue area. So a CIE 1976 uniform chromaticity diagram was defined using $(u', v')$ coordinates. The result is that comparisons in chromaticity in the $(u', v')$ space are less skewed from the bluer colors. This is called the CIELUV color space. L refers to the luminance of the colors used. A related color space is the CIELAB. The exact differences between these spaces are beyond the scope of this book. It is sufficient to know that one should be careful to know which color space an LED manufacturer is using. The software that

---

[1] http://assets.sylvania.com/assets/documents/faq0026-0999.f4172b60-cde8-4cb2-8a0a-37d3490d9e49.pdf
[2] http://www.citizen.co.jp/english/release/10/100302ecl.html

comes with spectroradiometers will crunch all the numbers and inform you of all the coordinates.

## COLOR RENDITION: HOW THE LIGHT LOOKS VERSUS HOW OBJECTS LOOK

While light sources of different color temperatures all look white, the objects they illuminate can look very different (see Fig. 3.13). The light sources all look white to the eye because the brain interprets it that way. Even the limited spectrum of the fluorescent light looks white because the eye is basically a three vector sensor. In other words, three monochromatic colors, red, green, and blue, can be mixed to arrive at any color. However, the perceived color of the object being illuminated depends on the spectral distribution of the light source. The object can reflect only the colors that are present in the light source. If a ruby of 650 nm wavelength is illuminated by a fluorescent light with a spike at 615 nm, then the reflected light will not be the same deep red as it will appear in daylight or incandescent light with significant power at 650 nm. Light from an LED source depends on the particular phosphor used, and varies by manufacturer. But in general LED sources have a broader spectrum than fluorescents and will have better color rendition. So a richly colored brown sofa will look better under incandescent and LED lights than under even a warm white fluorescent light.

The CIE measure for color rendition is the *Color Rendition Index (CRI)*. This is the industry standard for now but, again, it is far from perfect. The reference remains the black body radiator, as colors in daylight look natural. To find the CRI, a calculated comparison between the test light source and the reference is done with eight Munsell pastel colors (see Fig. 3.14, which lists the approximate Munsell). The difference on the color space is scaled by a factor to give 100 for the black body radiator and 50 for fluorescent lights. The General CRI, $R_a$ is the average of these

Figure 3.13    (a) Cool white fluorescent 4100 K, CRI 60; (b): Incandescent, 2800 K, CRI 100; (c): Reveal® incandescent 2800 K, CRI 78. (See color plate section.)

| Name | Appr. Munsell | Appearance under daylight | Swatch |
|------|---------------|---------------------------|--------|
| TCS01 | 7,5 R 6/4 | Light grayish red | |
| TCS02 | 5 Y 6/4 | Dark grayish yellow | |
| TCS03 | 5 GY 6/8 | Strong yellow green | |
| TCS04 | 2,5 G 6/6 | Moderate yellowish green | |
| TCS05 | 10 BG 6/4 | Light bluish green | |
| TCS06 | 5 PB 6/8 | Light blue | |
| TCS07 | 2,5 P 6/8 | Light violet | |
| TCS08 | 10 P 6/8 | Light reddish purple | |
| TCS09 | 4,5 R 4/13 | Strong red | |
| TCS10 | 5 Y 8/10 | Strong yellow | |
| TCS11 | 4,5 G 5/8 | Strong green | |
| TCS12 | 3 PB 3/11 | Strong blue | |
| TCS13 | 5 YR 8/4 | Light yellowish pink (skin) | |
| TCS14 | 5 GY 4/4 | Moderate olive green (leaf) | |

Figure 3.14 Approximate Munsell test color samples. (*Source:* http://en.wikipedia.org/wiki/Color_rendering_index under license http://creativecommons.org/licenses/by-sa/3.0/.) (See color plate section.)

eight numbers, R1–R8. This limited number of color choices limits the usefulness of CRI as a measure of color rendition. Also used are six other moderately saturated colors that represent colors common to the human experience such as flesh tone. These constitute the *Special CRI* that give additional information and are not used in the $R_a$ calculation.

There are obvious problems with this system. It can give high CRI to a source that does not render saturated colors. For example, some LEDs have a negative $R_9$ value, yet have an $R_a$ of 80. The reason for the negative value is that the delta on the color chart is so far off that the scaling factor doesn't really work. On the other hand, CRI can give a low value to a light source that enhances contrast, which gives preferred color rendering. For example, some RGB LED systems render the pastel colors more saturated than the reference incandescent and will yield a CRI in the 60 s even though it really ought to be greater than 100.

Typical fluorescent tubes for office use have CRI of 60. CFLs can be as high as 80. The better LEDs tend to be just above 80, as this seems to be the level where the color rendition starts looking natural. At 80, there will be slight color differences compared with incandescent sources, but the number is not so low as to render objects as a different color. There also exist CRI 90 LEDS where the red phosphor is increased to a longer wavelength to improve the $R_9$ value.

CRI and CCT do not give adequate guidance about how objects will look when illuminated. Although work is ongoing, it seems to us unlikely that any small set of numbers will ever be found that can do the job. As an example, look back at Figure 3.13, which shows the same scene with various CCTs and CRIs. It's not just that one looks worse than the other. Rather, the reds seem subdued on one; one seems brighter than the others; one "pops" more than the other, and so on.

## The Human Factor

There are many other factors involved in human perception of light and color. Color perception in the eye is the job of *cone receptors*. The cones are heavily concentrated in the center 1.5° of the retina called the *fovea*. This is one of the reasons why color perception changes when viewing a small sample of color versus viewing, for example, the color of the wall. On the other hand, *rods* sense only light intensity. Rods are everywhere except the fovea. In the center 1° of the eye, there are no rods. This is why when observing the night sky, where scotopic vision kicks in, there is a blind spot right in the middle of the retina and one must look slightly off center to see a faint star.

There is some variation in the human population in color perception. The physiological cause for this difference is a yellow spot over the center 4° of the retina. This spot can vary slightly in individuals and affect the color of the light reaching the retina. The cornea and the layers of cells in front of the retina can also vary in individuals and as they age. There may also be genetic differences in color perception between men and women (Jameson et al., 2001).

Lighting designers need to realize that there is a human factor in perceived lighting as well. The amount of light perceived by humans is *not* the same as the amount of light measured by a meter. The difference between the two is caused by eye dilation. As it gets darker, your pupil dilates, admitting more light. This causes the perceived light to be higher than the measured light. A typical conversion between the two is that the perceived light is the square root of the measured light. As an example, suppose that the measured light is 10% of full brightness. The perceived brightness is then approximately $\sqrt{0.1} = 0.32 = 32\%$. This factor is important in dimming lights. You need to have a wider dimming range than is suggested just by measuring lumens.

You should be aware of the possibility that the type of light produced by current LEDs may have some adverse effects on humans. The International Dark-Sky Association released a statement[3] that "Unfortunately, bluish light produces high levels of light pollution with significant environmental impact. These lights are known to increase glare and compromise human vision, especially in the aging eye. Short wavelength light also increases sky glow disproportionately. In addition, blue light has a greater tendency to affect living organisms through disruption of their biological processes that rely upon natural cycles of daylight and darkness, such as the circadian rhythm. . . . Circadian rhythms are controlled by light emitted within the dashed curve (see Fig. 3.15). The color of light emitted by a typical bluish-white 5500 K LED is depicted by the bold line. A large portion of light emitted by this light source falls outside of the human photopic vision range, and falls within the circadian rhythm curve."

On the other hand, there may also be benefits to the use of blue light as present in LEDs. Some studies have suggested that low levels of blue light can be effective at treating mood swings or low energy levels in some people. Red LED lights may be effective against skin aging. And the fact that scotopic vision has greatly higher

---

[3] "Blue light threatens animals and people," http://www.ledsmagazine.com/press/20192. But also see http://docs.darksky.org/Reports/IDA_Blue-Rich_Light_White_Paper051710.pdf and http://docs.darksky .org/SB/LED-SB-v3i1.pdf.

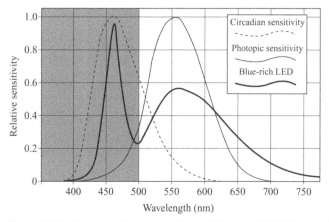

Figure 3.15   Circadian rhythm sensitivity. (*Source:* "Visibility, Environmental and Astronomical Issues Associated with Blue-Rich White Outdoor Lighting," May 2010, IDA. Image copyright of IDA.)

Figure 3.16   Identical gray boxes look different depending on their background.

sensitivity than photopic may mean that the light intensity required for street lighting may be lower with bluer light than might otherwise be supposed.

Color perception is also a psychological phenomenon (see Fig. 3.16). A gray color will look darker on a light gray background and look lighter on a dark gray background. The brain tries to compensate for the lighting condition. The light gray background signals to the brain that the illumination is very bright. Therefore, the gray color is actually darker than it would appear and so the brain interprets the gray color to be darker. In Figure 3.16, if you stare at the gray box on the dark background, you will train your brain to think that the lighting is dark. Over a period of a couple of seconds, you will notice the center gray square becoming brighter and seem to pop out more. This effect also happens with different colors. It is almost impossible to identify the real color of a sample when it is surrounded by different colors.

# CHAPTER 4

# PRACTICAL CHARACTERISTICS OF LEDs

## CURRENT, NOT VOLTAGE

The first thing to know about LEDs—and all diodes—is that they are current devices, not voltage devices. What does this mean? With a resistor, voltage and current are proportional. If you place a voltage across a resistor, a certain current will flow through it. By Ohm's law ($V = IR$, or volts equal ohms multiplied by amps), this current will be the voltage divided by the resistance, $I = V/R$. If you double the voltage, the current will double. Conversely, if you push a current through a resistor, it will develop a voltage across it. Again, by Ohm's law, this voltage will be the current times the resistance, $V = I \times R$.

But with diodes, Ohm's law doesn't apply. Voltage and current are not proportional. In fact, they are more like exponentially related. Specifically, the current through a diode (to first approximation) can be modeled by

$$I(V_f) = I_0 e^{k \times V_f},$$

while its inverse models the voltage through it:

$$V_f(I) = \frac{1}{k} \times \ln\left(\frac{I}{I_0}\right).$$

In these equations, $V_f$ is the voltage from anode to cathode of the diode, $I$ is the current through the diode, $k$ sets the scale for the voltage, and $I_0$ sets the scale for the current. $k$ and $I_0$ are constants. As typical values for white LEDs (at room temperature), we will take, $I_0 = 3.2\,\mu A$ and $k = 3.64/V$. These values are those for actual commercially available devices we will estimate looking at computer modeling of LEDs in Chapter 15.

Let's look at the equation for current for a moment. The thing to note is that the current through the diode is a very strong function of the voltage. Using the typical values for white LEDs as above, in order to put 2.80 V across the diode, we have to put 85 mA of current through it. Adding just a quarter volt, to 3.05 V, more than doubles the current, to 212 mA. And another quarter volt, to 3.30 V, again more than doubles the current, to 527 mA.

---

*Practical Lighting Design with LEDs*, Second Edition. Ron Lenk and Carol Lenk.
© 2017 by The Institute of Electrical and Electronics Engineers, Inc. Published 2017 by John Wiley & Sons, Inc.

What this means is that it isn't reasonable to try to control the diode's operation by controlling its voltage. Let's look at it the other way, with the second equation. When you put 100 mA of current through the diode, the forward voltage is 2.843 V. Doubling the current by adding 100 mA more increases the voltage by only 191 mV, less than 7%.

It is in this sense that we say that a diode, and an LED in particular, is a current device and not a voltage device. To a practical engineering first approximation, the forward voltage is always the same, regardless of how much current you put through it, so the performance of the device is determined by how much current you put through it.

This also means that, to the same approximation, the power into the diode is determined by the current. Since power is voltage times current, and voltage is constant, power is just proportional to current. As we'll see in the chapters on power supplies for LEDs, this is the way they are actually driven. The optical output of an LED is specified in lumens/watt. After deciding how many lumens are needed, this tells you how many watts are needed, and then this tells you how much current is needed. Power supplies for LEDs are typically designed to drive them with a constant current.

## FORWARD VOLTAGE

For easy estimates, the forward voltage of a diode is a constant. But the forward voltage does vary somewhat with current. And since each diode is different, where do the numbers come from? How can different diodes be compared?

Let's talk first about ordinary silicon diodes used as rectifiers. They have typical $V_f$ values ranging from 500 mV for small signal diodes to 1.2 V for large rectifiers (Schottky diodes are a different type of device). These numbers are a sort of practical guide to the performance of the diode. For a given amount of current, the diode with a lower forward voltage will dissipate less power. (Why wouldn't you always use the diode with the lowest forward voltage? Lower forward voltage usually means a physically larger device.)

Suppose you want to have a 500 mA current through your diode. If power dissipation during conduction is the most important thing, you could go to a database and assemble a list of all the 1 A diodes and see which has the lowest forward voltage at 500 mA. Easier at this level of approximation is just to look at forward voltage at the rated current of 1 A—a number that is always right on the front page of the datasheet. And indeed, this is how manufacturers and distributors have their online databases set up: forward voltage at rated current.

Now of course, nothing analog is really that simple. Forward voltage also depends on the temperature of the die, and this depends on how big the package is. The same diode in a bigger package will stay cooler, and thus have a higher forward voltage. Conversely, you can get a somewhat lower forward voltage by moving to a diode rated at a higher current. For example, a 3 A diode rated at the same forward voltage as a 1 A diode will have a lower forward voltage when operated at 1 A. But for a first cut, a simple spreadsheet lineup of $V_f$ will do.

The same principles apply to LEDs. For white and blue LEDs, typical forward voltages range from about 3.1–3.8 V. Yellow LEDs are somewhat higher, and red LEDs are down around 2.2 V. It's actually a little easier comparing forward voltages of LEDs than ordinary diodes. It's an industry standard to report the $V_f$ at a current of 350 mA, even for devices capable of carrying 1 A.

As an aside, note that the underlying reason that LEDs have much higher forward voltages than silicon diodes is that they aren't made of silicon. Their bandgap is different than silicon, that's how they generate light. And the reason there's such a range of voltages for white LEDs is that there's a number of different semiconductors being used in the industry now. Each has its own bandgap. One of the major areas of research for all LED die manufacturers is how to reduce the forward voltage of the device. Reducing this would increase the light output of the LED per watt.

## REVERSE BREAKDOWN

All diodes will conduct current when a voltage is applied from anode to cathode. All will also conduct if enough voltage is applied from cathode to anode—whether intentionally or not. Zener and avalanche diodes fall into the intentional category; this type of conduction is their main operation mode. You select one of these diodes based on what voltage you want conduction to occur at. The manufacturers hold the tolerance on this voltage to 5%, or even better.

Rectifier diodes and LEDs fall into the unintentional category. If they conduct in the reverse direction, there's an excellent chance that they've broken. Now with rectifier diodes, there's an easy solution. The reverse breakdown voltage is one of the parameters you select for when choosing the diode. Diodes of 200, 400, and 600 V are all commonly and inexpensively available, usually with very similar other characteristics. So if your diode breaks down, you can just replace it with a higher voltage part.

LEDs are not so convenient, unfortunately. Most of them have a reverse breakdown voltage of only 5 V. This very low voltage can present a serious problem in a practical circuit. Even with several LEDs in series, the breakdown is still only tens of volts. There isn't any room for error. Any glitch or noise in the control loop of the power supply, or a hiccup in the AC line voltage, may be enough to momentarily generate breakdown. It doesn't take long—diodes can be broken in microseconds under the right conditions.

One obvious solution to try is to put a regular rectifier diode in series with the LEDs, as shown in Figure 4.1. This prevents any reverse current from flowing through

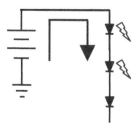

Figure 4.1 Reverse bias protection.

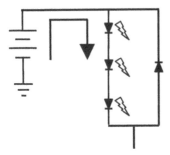

Figure 4.2   LEDs with reverse bias protection.

the LEDs. Unfortunately, it also dissipates power when the LEDs are operating normally. If the LEDs are being driven at 400 mA, the forward voltage of the blocking diode of about 1 V will dissipate about 400 mW of power all the time. This may be unacceptably high, either for thermal or for efficiency reasons. Thermally, it may be difficult to get rid of the excess heat of 400 mW that isn't doing anything for you most of the time. For efficiency on small light sources, 400 mW may be a significant fraction of the total power into the device.

The other choice is to put a diode in antiparallel with the LEDs, as shown in Figure 4.2. Now if there should happen to be a voltage applied from cathode to anode of the LEDs, it is clamped to the forward voltage of the diode, about 1 V. This is absolutely safe for the LEDs. It also doesn't dissipate any power in normal operation, as the diode is reverse blocking when the LEDs are conducting normally.

As long as we're on the topic of LED current, it's worth a quick look at ratings. Being semiconductors, all LEDs have an absolute maximum current rating from the manufacturer. These are typically 700 mA or 1 A, but there are many others. With a normal diode, you know that the current rating is really for steady-state operation. For a pulse, you can usually have much more current. For example, the 1N4973 is a 1 A diode, but it can handle a 30 A surge for half a line cycle, 8.3 ms. So what about LEDs? The answer is different for each manufacturer and for each type of LED.

Some white LEDs can be driven above their rated current by a large amount. Their phosphors saturate, and the light starts to turn blue, but if this doesn't bother you, they can take quite a bit more current. In general, current generation LEDs seem to have plenty of bond wires to carry a lot of current. So the real limitation is whether the die can take more current without burning up or developing hot spots. This is primarily a thermal question until you get to very high currents. And as we'll see in the next section, very high currents are bad for efficacy, and so you probably don't want them anyway. But a little bit of overdrive is probably okay.

## NOT EFFICIENCY — EFFICACY!

Efficiency usually refers to electrical conversion. You put a certain amount of energy into a power supply; you get a certain less amount out. The electrical efficiency is defined to be the output power divided by the input power. It is usually expressed as a percentage. Since there are always some losses, it is a number always less than 100 and not less than 0.

Things are not so straightforward for LEDs. You put power in, but what you want out is not power, but light. Now light, just like electricity, *can* be measured in watts. And indeed, deep blue LEDs' light output is measured in watts. Efficiency is then straightforward. It is the output power in watts divided by the input power in watts. As an example, the LXML-PR01-0275 from Lumileds is a royal-blue LED (455 nm). It has an optical output power of 275 mW, and an input power of 350 mA × 3.15 V = 1.103 W. Its efficiency is thus 0.275 W/1.103 W = 24.9%.

As discussed in Chapter 3, humans have very little vision at 455 nm. Nobody would try to use such a light to see by. It therefore makes sense to specify its output in watts. Other colors, including white, are different. For example, for traffic lights we will want to know how bright red, yellow, and green appear to people. Similarly, for light bulbs using white LEDs, we want to know how bright people perceive them to be, not how much optical power is coming out. So for these nonblue LEDs, watts aren't used, lumens are. And so efficiency isn't the right term either. Instead, we use efficacy. (Although the authors don't know this for a fact, it is worth observing that the lumens output of a 455 nm LED is very low, because humans don't perceive that color well. As a result, the lumens output of such a device would be very low. Perhaps this would look bad on a datasheet compared to devices emitting light that people see?)

White and other LEDs' efficacy is measured in lumens per watt (Lm/W). Lumens are used because this reflects the amount of light that humans perceive, as described in Chapter 3. Now this has some strange effects. To start, the human eye response peaks in the green part of the visible spectrum. This means, for example, that for a red and a green LED with the same efficacy, the red must be emitting a lot more light than the green. You thus can't directly compare two different LEDs based on efficacy unless they are the same color. Even two quite similar colors—say greenish yellow and yellowish green—aren't directly comparable.

Additionally, LEDs don't emit exactly one wavelength. They're not lasers. For example, Lumileds's LXML-PM01-0040, a green LED, has the dominant wavelength specified to be between 520 and 550 nm. Additionally, the spectral half-width is 30 nm. So the light is somewhere between 490 and 580 nm. Again, human eye response peaks at 555 nm, and so the efficacy of these devices will vary widely, even when in terms of optical power they are similar.

When you get to mixtures of colors, the confusion is even worse. And white light is by definition a mixture of all the visible colors. As discussed in Chapter 1, white LEDs are made by adding two or three different colors together. So the efficacy of such a device depends very strongly on the exact mixture of colors, on the exact brightness of each color in the mix, and on exactly where in the spectrum each color is emitting.

Finally, white LED vendors consider efficacy to be the parameter on which they are most significantly competing. So, they tend to play games with the spectral composition of white LEDs. Shifting closer to green will increase the efficacy, so shifting the CCT from 2750 to 2900 K produces a better number.

Efficacy as specified in datasheets is measured at nominal current, typically 350 mA. As you change the current, the efficacy also changes. Now when you decrease the current, there isn't a huge change, maybe a 10% increase if you go to 100 mA. So, practically, you can use the datasheet number. But if you increase the current, the efficacy starts to decrease substantially. So if you're concerned with efficacy, beware. Running an LED at more than its nominal current does produce more than nominal light, but the efficacy suffers.

Let's take a look at how this works out practically. Lumileds' LXML-PWW1-0060 has a guaranteed minimum output of 60 lumens at 350 mA. (At 25 °C—we'll delay talking about temperature effects till Chapter 5.) To know the efficacy, however, you need to also know the power level, not just the current. We find in the datasheet that the nominal forward voltage of the device at 350 mA (and—again—at 25 °C) is 3.15 V. Thus, its nominal efficacy is 60 lumens/(350 mA × 3.15 V) = 54.4 Lm/W. Note that the efficacy is *not* 60 Lm/W—some sales people have been known to confuse lumens and lumens per watt.

Next let's figure out how the efficacy changes with drive current. The temptation is to look at the curve (Fig. 4.3) "relative luminous flux" and say, "The light output increases this amount, and therefore so does the efficacy." But that's incorrect on two scores. The efficacy depends not just on the light output, but also on the drive current and the forward voltage. And while the flux is increasing, the drive current is also increasing, and so is the forward voltage (Fig. 4.4). The efficacy as a function of drive current is not shown in the datasheet!

We use the data in the two figures provided by the datasheet to construct the curve of efficacy versus drive current in Figure 4.5. You should note that above the nominal drive current, the efficacy of the LED starts to decrease quite a bit. To get a feel for the effect of this, suppose you want to get 100 lumens. At 350 mA, you need 100 Lm/(54.4 Lm/W) = 1.84 W. You could do this by running two LEDs both at something close to 350 mA. Suppose instead you do this with a single device. Since you need something more like 700 mA, the efficacy is down to 42 Lm/W—at which you need 100 Lm/(42 Lm/W) = 2.38 W. You're spending considerably more power (29% more) to get the same light output, and perhaps falling below specification on

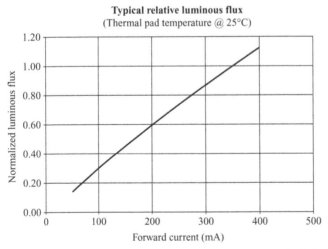

Figure 4.3   Light output as a function of current. (*Source:* Technical Datasheet DS56, Power Light Source Luxeon Rebel, Philips Lumileds Lighting Company, 2007.)

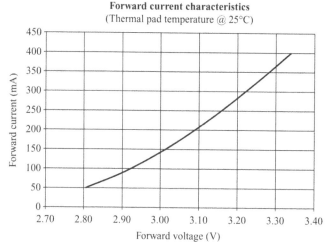

Figure 4.4   Forward voltage as a function of current. (*Source:* Technical Datasheet DS56, Power Light Source Luxeon Rebel, Philips Lumileds Lighting Company, 2007.)

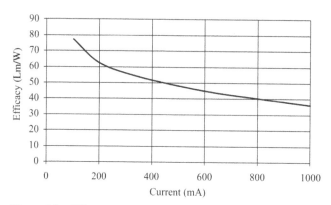

Figure 4.5   Efficacy versus drive current.

required efficacy of the light source. Of course on the other hand, you're paying less money for one LED instead of two. It's one of those trade-offs engineers always face. We'll see more of this in Chapter 9.

## LED OPTICAL SPECTRA

Since the goal of an LED is to produce light, its optical spectrum is a key performance parameter. Let's take a look at a typical spectral curve for a warm-white LED, shown in Figure 4.6. Along the *x*-axis is wavelength in nanometer and along the *y*-axis is relative spectral power. Note that the manufacturer doesn't label units on the *y*-axis: All you can say are statements such as "the output at 675 nm is half that at 600 nm."

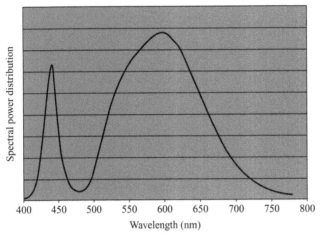

Figure 4.6   Light output as a function of wavelength. (*Source:* Technical Datasheet DS56, Power Light Source Luxeon Rebel, Philips Lumileds Lighting Co., 2007.)

A few things stand out. While the bulk of the curve resembles the human eye response curve, one part is quite different. There's a big spike in the blue region. The origin of the spike is obvious. This is the native emission of the die, and the spike represents the portion of that emission that is not being converted by the phosphors. This doesn't really hurt the color of the LED, since humans see very poorly in blue. But it does represent one of the areas in which LED performance can be improved. Converting all of the blue light to the more useful part of the spectrum would increase efficacy.

The other noticeable thing about the curve in Figure 4.6 is how quickly it tails off in the red. Getting a little more red up, there is the difference between 2800 and 3500 K CCT. Even more importantly, it's also the difference between reddish objects looking brown versus looking red. So this part of the curve gives some idea of how reddish red objects will look. Unfortunately, making deep red phosphors is currently the most challenging task for LED manufacturers. This is why many present-day white LEDs have poor CRI. They drop out in the reds (particularly R9; see Fig. 4.7).

The optical spectrum of the LED tells you everything about its light. But there's no practical way to look at the spectrum and decide if the light is a reasonable color for a given application. So as has been discussed in Chapter 3, what people have done is to make up a set of numbers that characterize important features of the spectrum. These are numbers such as CCT, CRI, and $(x, y)$. As discussed, these have pluses and minuses. The CCT gives some idea of how "cold" the light is, at least for white light. The CRI gives some idea of how well colors are reproduced by the light—to the extent that the spectrum is blackbody. The various characterizing numbers are useful because they give *some* idea about what the light looks like. But the real reason they're useful is because they are what LED manufacturers can characterize. Ultimately, there's no substitute for building your design with the actual LEDs and trying them out.

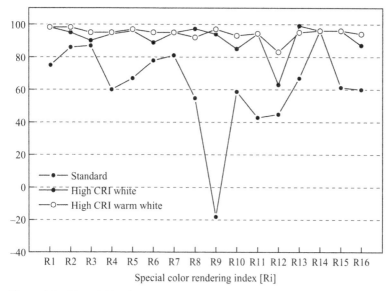

Figure 4.7    Many LEDs have poor R9. (*Source:* http://www.yegopto.co.uk/LightingLEDs/ CRI_Seoul_Semi.)

Despite the limitations of the characterizing numbers, we can discuss how they vary. Variation should not be a surprise. If you drive too much current into a white LED, it turns blue. The phosphor has saturated, and the blue light of the LED comes through without being converted. In general, phosphors are complicated molecules, and how much and how efficiently they capture and convert blue light must depend to some extent on things such as drive current. RGB LEDs are even more susceptible to variation with current, since their efficacy varies with current.

The color coordinates of white LEDs vary with drive current. Look at Figure 4.8, from an Intematix datasheet. As drive current increases, both $x$ and $y$ increase (400 mA is maximum rated drive current for this device). This variation, while not huge, is enough to change the appearance of the LEDs. And even with this information, we keep noting that the $x$ and $y$ coordinates alone aren't enough to specify what it's going to look like. Most manufacturers don't specify even this amount of information, so if you plan to run the LEDs at anything other than their nominal current, you have to build and measure the system to find out what the light will look like.

One more optical characteristic to consider is angle of emission. The light doesn't come out the back of the LED, because it's mounted on a package there. In some LEDs, the light is emitted as you would expect, from the front of the die. But other lights are emitted perpendicular to the die. Further, the light emitted from the LED die usually goes through some optics in the package, which further modifies the light distribution. The result is that the light output is not the same intensity at all angles; there is more light in some directions than others.

There are three common light emission distributions for LEDs: Lambertian, batwing, and side emitting. Lambertian is the most common. Looking at Figure 4.9,

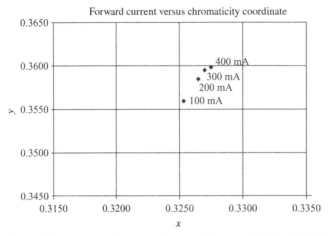

Figure 4.8 $(x, y)$ as a function of current. (*Source:* C6060-16014-CW/NW Datasheet, Intematix Technology Center Corp., 3/2008.)

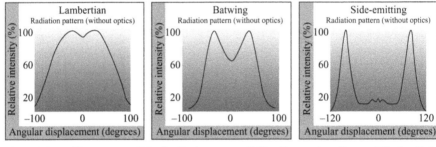

Figure 4.9 Different output light distributions are available. (*Source:* http://www.philipslumileds .com/technology/radiationpatterns.cfm.)

the fundamental difference is clear. Lambertian is primarily forward-directed light. Batwing is also forward-directed, but has an annulus of light, with a dip in the middle (remember that the diagrams need to be rotated out of the plane of the paper). Finally, as the name indicates, side-emitting LEDs emit most of the light out of the sides, and very little is forward-directed.

Most manufacturers specify a radiation pattern, without being too specific about what this exactly means. But there are at least two important measures of angular light emission. Lumileds specifies both a viewing angle and an included angle. The included angle is probably the more important of the two, and is probably what other manufacturers mean when they specify an emission angle. The included angle is the (circular) angle in which 90% of the light is emitted. What this means practically is that if the angle from the LED doesn't run into anything up to this angle, then 90% of the light emitted by the LED will actually escape from the device. Lumileds defines this to be "total angle," so that their 160° specification corresponds to ±80°. For other manufacturers 120° is more common, and this presumably is also the total angle.

Viewing angle is here the angle at which the brightness is half what it is at the maximum. (This is what was called "beam angle" in Chapter 3.) For most applications, this number is less important. Note the fundamental difference between the two angles. Viewing angle gives you the same information as the emission curve, the angle at which relative intensity is half. Included angle is how much of the total curve is included—the integral of the area under the curve. If the angle from which you extract light is anything smaller than the included angle, you have to look at the emission curve to figure out how much light you're extracting. If it's bigger, then you know you're getting more than 90%.

## OVERDRIVING LEDs

We just said that driving LEDs at currents higher than their specification results in color shifts. And of course it also means the LEDs get hotter from the additional current, plus a small additional factor from the forward voltage rising with additional current. A hotter LED has a shorter lifetime. Given all these negatives, why would you want to overdrive an LED?

The main driver is usually cost (although size can be a factor, too). Five LEDs cost 25% more than four LEDs. If instead I can drive each of the four LEDs 25% harder, I save the cost of that additional LED. So what should you do? Color shifts are usually not that important for today's market, so it still comes down to LED temperature. If you can hold the appropriate die or case temperature down to a number that gives you adequate lifetime, go ahead and do it. We think this is the important factor as long as the absolute maximum current rating of the device isn't exceeded.

And if that extra 25% would exceed the absolute maximum, then you have to consult with the vendor. Of course, they will tell you that you can't do it, that there is no guaranteed performance beyond that rating, and that your system will fail horribly. This is certainly possible, but some digging may reveal which part of the LED is the first to go. The bond wires? Die hot spots? Phosphor saturation? There's failing such that the part no longer works, but there is also failing where the part no longer exactly meets all of its specifications. Finding out which one the absolute maximum current rating is based on may allow you some leeway to exceed even this rating, at least by a bit.

## KEY DATASHEET PARAMETERS

The previous sections have talked about quite a number of parameters for LEDs, as well as their temperature variations. It would be nice if LEDs had just a single parameter to characterize their suitability for an application. That way you could just say, "This LED is a #8, and I need a #9." Unfortunately, LEDs are deeply embedded in the analog world. There are half a dozen parameters that are important, and no one device is going to exactly meet all of your criteria.

Let's try to summarize the most important criteria in a practical sense. The first thing to do is to select the CCT (or color, if you're not building a white light source). This usually eliminates 2/3 of the choices. CRI is less important, because most devices are about the same unless you want an unusually high CRI.

After CCT, the next criterion is usually the efficacy at operating temperature. We have to re-emphasize that this should not be at the 25 °C manufacturers specify, because different devices' optical outputs change differently with temperature. Take the light output at 25 °C, multiply this by the decrease in light at 85 °C (usually a good starting point), and divide by the current specified and by the forward voltage, again compensated for temperature:

$$\eta(85\ °C) = \frac{\text{Lumens}(25\ °C) \times \text{relative light}(85\ °C/25\ °C)}{\text{Current} \times V_f \times \text{relative } V_f(85\ °C/25\ °C)}.$$

You want to select a device with a minimum efficacy given by how much light you want to get out and how much heat you can dissipate. Usually, but not always, going to higher efficacy devices beyond this increases the cost.

# BINNING

Realizing how much the variation in some of these parameters influences performance, manufacturers offer binning. Binning means that the manufacturer offers the same part with differences in the key parameters. The buyer selects which bin the parameters should be in. This allows some degree of control over the parameters. The parameters are otherwise usually fairly broadly defined. Part of the reason for this is that it's hard to control the parameters over different production runs; the vendors want to ensure they sell everything they make.

As an example, all manufacturers offer CCT binning. If you look at "neutral-white" on Lumileds' datasheet, it means the CCT is between 3500 and 4500 K. But this is such a broad range that the parts actually available are binned. Figure 4.10 shows how this neutral-white space is binned. There are limits on both $x$ and $y$, resulting in 12 color bins.

There are more bins offered above the Planck line than below. Four of the bins straddle the Planck line; half are above and half below. This means that LEDs from the same bin may produce some greenish and some reddish white light. It should be noted again that this is an ANSI standard (ANSI C78.377), and so the problem is not with Lumileds: All the vendors are specifying bins this way.

Note that even within a single bin there can be variation in the color of the bin, because, of course, a single $(x, y)$ coordinate cannot be produced. The key question is whether this unit to unit variation is perceptible by customers. The generic answer you get from manufacturers is "no," within a single bin you can't tell the difference in color. However, the authors believe that variation is noticeable, at least to some portions of the population. Not everyone sees color the same, and some are more sensitive to variation than others (see Chapter 3). This viewpoint is backed up by the

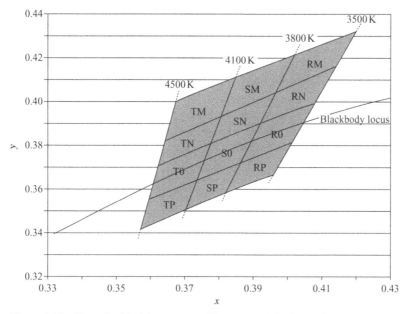

Figure 4.10   Neutral-white bin structure. (*Source:* Technical Datasheet DS56, Power Light Source Luxeon Rebel, Philips Lumileds Lighting Company, 2007.)

fact that some manufacturers are going to substantially smaller bin sizes. In any case, this is something for the marketing department to work on. Since there are differences in color perception among humans, be sure you are not the sole judge of whether the binning is good enough!

We note that there's some trickery involved with this color binning. Certain manufacturers specify all of these bins, but then you can't buy just one of them. For example, they may require you to accept any one of four or even six adjacent color bins. Every time you get a shipment, its bin will be labeled. But you may get stuck with bins you can't use. Ensure you talk with your vendor about what bins they actually ship versus what the datasheet says.

Other binned parameters may include forward voltage and brightness. Brightness is always binned, because this is what drives the price. Brighter lights cost more. But it may turn out to be more cost-effective to buy a lower brightness bin, while paying for a lower forward voltage bin. If what matters is efficacy, both the light output and the power input matter. Again, you have to talk with the vendor to understand how the pricing structure is set up.

## THE TOLERANCE GAME

You've selected a part based on consideration of its luminous output and color (and of course cost). Its forward voltage is whatever it is. Now you look at the datasheet to find out minimum efficacy at the current you're going to drive it, and figure out how many of these LEDs you need. (We'll have detailed examples of how you actually do this in Chapters 10–13.) Everything's ready to go, right? Wrong! Hidden inside the datasheet is a spec that can kill your calculation. It's hidden away as a footnote and is easy to miss.

This hidden menace is the measurement tolerance on the light output. Of course all parts have a tolerance on their light output, or on their efficacy. You've already accounted for this by calculating light output from your device at minimum guaranteed LED light output. The problem is rather that the manufacturers don't really guarantee a minimum light output. They guarantee a minimum light output, *given their measurement accuracy*. This accuracy is usually specified as ±7%, or ±10%. So a datasheet that says the LED has an output of 90–100 lumens could have an output anywhere from 81 to 110 lumens.

Now, by an interesting coincidence, whenever the authors have measured lumens from various manufacturers' LEDs using their own calibrated equipment, the parts are always at the bottom of their bin and at the bottom of their measurement tolerance! So the part that you might think was supposed to be a typical 95 lumens and a minimum of 90 lumens is almost always 81 lumens. Now if this were really a tolerance problem in their optical measurement, you would expect a distribution of brightness. Since number of lumens is an important factor in marketing, you have to wonder about truth in advertising. In any case, if you need to guarantee the brightness of your device, you need to account for the measurement accuracy as well.

CHAPTER *5*

# PRACTICAL THERMAL PERFORMANCE OF LEDs

IN CHAPTER 4, we looked at a variety of important parameters of LEDs, both electrical and optical. All of these parameters are affected by temperature and temperature-related aging. And the dependencies need some detailed explanations in order to be useful. For that reason, we've separated out the discussion of thermal performance of LEDs into this chapter. Measuring the temperature of the LED is yet another concern. While there is some information on that in this chapter, you should look at Chapter 14 for a detailed discussion of experimental technique for this tricky measurement.

## MECHANISMS BEHIND THERMAL SHIFTS

You'll recall that a white LED consists of a number of different components. There's the die itself emitting blue light. There's a phosphor that converts part of the blue light into white light. There's the silicone encapsulant to protect the die and the phosphor mechanically. And then there's the package into which the whole thing is mounted. All of these components potentially contribute to thermal performance of the LED. And they each contribute differently again to thermal aging.

Let's start with the die. Since there are many different materials used for die, some proprietary, manufacturers haven't revealed information about their thermal performance. (Thermal performance specified in datasheets is for the LED as a whole, not the individual components of the LED.) But we can take some educated guesses based on what we know about silicon. There is certainly an absolute maximum temperature beyond which the die simply fails. There is probably also a thermal run-away temperature, at which certain parts of the die become hotter than other parts (hot spots) and fail, causing the whole die to fail. Finally, it seems likely that the wavelength of the emitted blue light shifts with thermal aging. The shift is probably fairly small, but then the phosphors absorb light in a very tight band of wavelengths. A small shift in the emission wavelength may be enough to significantly affect the

*Practical Lighting Design with LEDs*, Second Edition. Ron Lenk and Carol Lenk.
© 2017 by The Institute of Electrical and Electronics Engineers, Inc. Published 2017 by John Wiley & Sons, Inc.

ability of the phosphor to absorb the light, resulting in a decrease of efficacy with thermal aging.

The phosphors are an obvious source of temperature effects. They are tuned to absorb at a certain wavelength, and this obviously must vary with thermal excitation of the molecules. They also degrade with temperature, red-emitting phosphors being notorious in this regard. Fixing this is a major research effort for phosphor companies.

The encapsulant is supposed to be mechanically strong, and at the same time optically clear. The latter is the problem. At first, epoxy was used. It has a number of desirable properties, including being inexpensive. But it turns yellow with age (especially fast-cure types). It was then replaced with silicone, now universally used for high-brightness LEDs. Silicone is better than epoxy. But it is still a polymer, and it too eventually turns yellowish with heat and time. This affects the color of the light being emitted, and reduces the efficacy of the LED (yellowish appearance means that other colors are being absorbed). Finding better formulations is a major effort for encapsulant companies.

In silicon, the package isn't usually a source of problems. But for LEDs, it can be. If the package is not perfectly reflective (and nothing is), then it is partially absorbing the light, reducing efficacy. And the package also has to survive high temperature for tens of thousands of hours. In some cases, the package gradually yellows, again affecting absorption and color emission. Cree (2010) says ". . . the primary mode of degradation for HB LEDs is the package itself." Here again is an engineering trade-off, cost of the package versus thermal aging performance.

## ELECTRICAL BEHAVIOR OF LEDs WITH TEMPERATURE

The most immediately noticeable performance change of LEDs with temperature is that their forward voltage drops. This is a die phenomenon, and so varies from manufacturer to manufacturer. Datasheets specify the typical decrease of forward voltage, which runs between $-2$ and $-4$ mV/°C, but they generally don't show a curve of $V_f$ versus temperature.

What does this mean practically? If you run the LED at constant current, as the device warms up the $V_f$ goes down. Thus, the power into the LED also goes down with temperature. And this in turn means that the light output also goes down. A back-of-the-envelope calculation gives a feel for what this means. Suppose the 25 °C forward voltage is 3.6 V, and also suppose to make the calculation easy that the $dV_f/dT$ is $-3.6$ mV/°C. If the temperature goes up from 100 to 125 °C, then the forward voltage drops by $-3.6$ mV/°C $\times$ 100 °C $= -360$ mV, to 3.24 V. Since this drop is 10% of the room temperature forward voltage, the power into the device is also reduced by 10%. Even if the efficacy of the LED were unaffected by power, which isn't true, the light output would be down 10% at temperature just because the input power is down.

Now, a way to ensure that the light output is not affected by this $V_f$ drop is to run the LED at constant power. We could measure the voltage across the LED, and increase the current as the voltage dropped to maintain constant power. Suppose we were putting 700 mA into the LED at room temperature, a power of 3.6 V $\times$ 0.7 A $= 2.52$ W.

At 125 °C, the voltage has dropped to 3.24 V, so we need to increase the current to 2.52 W/3.24 V = 778 mA, an 11% increase.

That the datasheet shows just a typical $dV_f/dT$ might make us suspicious. Is there actually a nonlinear curve of $V_f$ versus temperature? We've done a few measurements that suggest that the curve is more or less linear. Indeed, measuring the forward voltage is one way of measuring the temperature of the die. But experience with silicon suggests that this may be a good approximation over some range of common operating temperatures, and may vary beyond those limits. Testing the actual performance of the specific LED you intend to use at its actual operating temperature is the only way to be sure of the temperature's effect on the LED.

## OPTICAL BEHAVIOR OF LEDs WITH TEMPERATURE

Thermal effects on electrical performance of LEDs depend only on the die. Optical effects involve all the components of the LED, and so are more numerous. To start with the most commonly discussed effect, the brightness and efficacy of an LED decrease with increasing temperature. Now, if you look at a typical datasheet, what is shown is a curve of (relative) brightness versus temperature at a constant current (see Fig. 5.1). Since constant current is the common way to drive LEDs, this is a useful curve. But the reality is that this mixes together two effects. At temperature, the efficacy decreases, but the constant current means that the input power is also decreasing. So the datasheet shows you the brightness decrease, *not* the efficacy decrease.

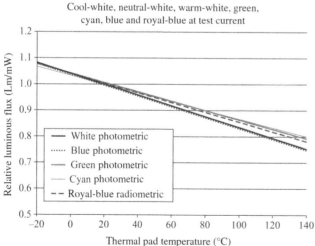

Figure 5.1    Brightness as a function of temperature. (*Source:* "Technical Datasheet DS56, Power Light Source Luxeon Rebel," Philips Lumileds Lighting Company, 2007.)

Let's find the efficacy as a function of temperature. The graph shows that from 25 to 125 °C at constant current, the light drops from 1.00 to 0.78. (Note that this is flux, not efficacy. The "Lm/mW" on the $y$-axis means "lumens or mW," and is because "royal blue" is measured in watts, and all the other colors are measured in lumens.) Now, this is a 100 °C delta. The datasheet also shows that the $dV_f/dT = -3\,\text{mV/°C}$, so the forward voltage drops from a typical value of 3.15 to 2.85 V. This means the input power drops from $3.15\,\text{V} \times 350\,\text{mA} = 1.103\,\text{W}$ to $2.85\,\text{V} \times 350\,\text{mA} = 0.998\,\text{W}$. This is a drop to 90% of input power. Efficacy thus has dropped from 1.00 to $0.78/0.90 = 0.86$, or, in other words, efficacy has dropped 14%. The other 8% of the light drop is, again, due to the constant current drive circuit producing less power for the LED at high temperature.

There are other effects of temperature on the light beyond the efficacy drop. The color of the light changes to some extent with changes in temperature. This is usually not specified on the datasheet, although once in a while you will see a graph of $(x, y)$ versus drive current. This seems to be due to small shifts in the phosphor emission with temperature. Since we know that red is the most susceptible to these changes, we may expect there to be shifts in both the CCT and the CRI. Since rarely the data are available on this, you have to go into the lab and measure it, and then decide whether it is a significant factor for your application.

## OTHER PERFORMANCE SHIFTS WITH TEMPERATURE

We haven't said anything yet about absolute maximum temperature ratings. Some LEDs are rated at 150 °C, others at 85 °C. This is probably due to a phosphor degradation effect. Our guess is that lifetime is the driver of this specification. For a variety of reasons, datasheets must specify that the LED drops to 70% of initial brightness after 50,000 h of operation (see the next section for further discussion of this). The only variable that can be changed is the temperature. By specifying a lower temperature, phosphor degradation is decreased, and so the time to 70% light is lengthened. We suspect that the effect is not all that dramatic. If the 70% point is after 40,000 instead of 50,000 h, are you going to reject the part? So it's not absolutely necessary to keep the LED below 85 °C. Indeed, it's hard to run any decent power through an LED and keep it this cool. You need to talk directly with the manufacturer and see what data they have on lifetime at higher temperatures.

There's also some subtlety in reading this specification. Exactly what temperature is it that is supposed to be kept below absolute maximum? Some datasheets say it is the die, others the pads. There's a big difference between the two. If you're putting 3 W into an LED with a thermal resistance from junction to pad of 10 K/W, that's a temperature difference of 30 °C, so when your pad is at 85 °C, your die is at 115 °C! Be sure you know at which point the temperature is measured for the lifetime rating.

One more performance consideration is the effect of pulsing current into the LED. Pulse-width modulation (PWM) of the current into the LED is a preferred method of dimming. This is because the current into the LED does not vary, but the amount of time the light is on does. If you dim the LED by reducing its current, there are some color shift issues, as described earlier. Now if you pulse the LED at high

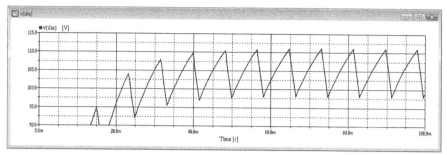

Figure 5.2    LED temperature profile for parameters given in the text.

frequency, the device reaches a temperature dependent on the average power, as you would expect. But if the pulses are slow, there may be time between pulses for the die to cool. If even part of the time the temperature is lower, this may have a significantly positive impact on the 70% life of the LED.

What sets fast against slow degradation is the thermal time constant. As we discuss in Chapter 6, thermals are exactly analogous to RC networks. And so in addition to thermal resistance, there is also thermal capacitance. The two together form a thermal time constant, which is how long the device takes to heat up when power is applied, and how long it takes to cool down when power is removed.

The thermal time constant of a typical LED is around 10 ms.[1] This means that after power is applied, it takes 10 ms to reach 63% of the temperature difference between start and end. Suppose you are putting pulses of current into an LED at 120 Hz (say from a dimmer circuit). Suppose also that its on-state duty cycle is 80%. We'll assume that when the LED is on, the power level is such that in steady state the die would reach 125 °C. Look at Figure 5.2 to see a simulation of the LED temperature profile with these parameters. The peak temperature the LED reaches is 111 °C. So the lifetime of the LED is considerably longer than it would be at 125 °C. In fact, since 111 °C is the peak temperature, the lifetime is longer than if the LED were steady state at 111 °C. But since it is complicated to figure out the actual temperature(s) at which the LED is operating, it is practical to use the lifetime at 111 °C.

## LED LIFETIME: LUMEN DEGRADATION

The previous sections have talked about temperature effects that can be quickly measured in a lab. Now we turn to very long time thermal effects. As currently defined, an LED lamp is said to have a lifetime that is equal to the time required for half of the lamps to get to 70% of their initial light output (L70/B50). For better or worse, LED lifetime is defined the same way.

The first thing to observe is that this is difficult data to collect. Fifty thousand hours is about 8 years of continuous operation. With new generations of LEDs coming

---

[1] Inferred from Figure 4 of X. Poppe, "When Designing with Power LEDs, Consider Their Real Thermal Resistance", *LED Professional Review*, Nov/Dec 2009, p. 41. This number is order-of-magnitude confirmed by the standard pulse test for determining 25 °C efficacy, which is 25 ms long.

out every 6 months, there isn't time to measure life before the part is obsolete. So lifetime measurement has turned to extrapolation from short-time data.

In a very general way, we expect everything to age on a logarithmic time scale. If a parameter decreases by 5% in 1000 h, we expect that in the next 9000 h it will decrease a further 5%. Unfortunately, this Arrhenius law applies to a single aging mechanism. In an LED, we've identified at least four independent aging mechanisms: the die, the phosphor, the encapsulant, and the package. For a given device, there are potentially at least four different logarithmic aging time constants.

As an example of this, consider the odd case of the brightening light bulb. Some LEDs have had light curves showing that there is an initial *increase* in light output at constant current and temperature over the first 1000 h of operation. This appears not to be just a fluke measurement. Apparently some of the light bulbs the DOE tested have also gotten brighter in the first 1000 h. After that, they started their decline as expected.

In another case, the manufacturer found that there were two different log slopes. During the first 1000 h, the light decreased relatively rapidly. After that, the log slope decreased, and the light decreased very, very slowly. The conclusion here is that those various different aging mechanisms are interacting. That interaction is really what makes this problem so hard.

This problem is so trying that the Illuminating Engineering Society (IES), on which the government relies, gave up on it. LM-80, "Measuring Lumen Maintenance of LED Light Sources," was intended, as the name says, to define how well LED lights maintain their light with age. There was a great deal of committee work on the subject, from many of the manufacturers of LEDs. The conclusion was that there was no general way of estimating the 70% lifetime. The only way to know the lifetime is to actually measure it.

There's an additional complication in measuring the L70/B50 time. The definition says that this is the time at which half of the LEDs measured have reached 70% of their initial light output. So that's the mean time. But there's no specification of the standard deviation. Suppose that 48% of the sample fails after the first 2000 h, and then there aren't any more failures for 48,000 h. The L70/B50 time is 50,000 h, but that seems incorrect. Since no one knows how to extrapolate the data yet, this isn't a current problem, but someday it will have to be addressed.

## LED LIFETIME: CATASTROPHIC FAILURE

The lifetime of LEDs is treated differently from that of other bulbs. When the lifetime of an incandescent bulb is specified to be 1000 h, this means that after 1000 h half of a sample of such bulbs is burned out. Their filaments are broken and they produce no light whatsoever. Similarly, for a fluorescent bulb, when their lifetime is specified to be 8000 h (even though this number appears to be inflated), it means that after 8000 h half of a sample of such bulbs is burned out. Their filaments are broken or their vacuum is lost.

LED lifetime is different. With industry agreement, the government has mandated that the "lifetime" of an LED light is the time after which half of a sample of such bulbs has lost 30% of its initial light. The LED light bulbs are not burned out, they are just a little bit dimmer. Their lifetime to 50% failure is large, probably mostly set by the driver circuitry in the bulb. The LED itself has a lifetime that is probably enormous, perhaps hundreds of thousands of hours. This lifetime is set by catastrophic failures, such as bond wires breaking or lightning strikes. For example, one manufacturer's test showed that after 50,000 h, a sample of 68 devices had only one real failure.

How did such an inequality of treatment arise? Fluorescent bulbs have been very carefully engineered. It turns out that at roughly the time when they fail, their light output is about 70% of initial. And 70% is significant because it's the level at which people can start to tell that a bulb is dimmer. It thus appears that the problem is not with the definition, but with the LEDs' failure mode. Here's a proposal. Can manufacturers find one of the construction parameters to adjust such that L70 = B50? That would bring LEDs into parity with the other light sources. And it would be even better if that adjustment would result in a cost reduction.

## PARALLELING LEDs

One thing that is rarely done is to put diodes in parallel. As LEDs are diodes, we should expect the same to be true for them. Let's take a quick look at why. If LEDs were absolutely identical, paralleling them would work. For example, in a Spice simulation, each LED has exactly the same forward voltage versus current. If you attach two of these simulated LEDs to a current source, each will take exactly half of the current. The reason this works is the negative feedback of the forward voltage. If the current in one were higher than in the other, it would have a higher forward voltage. But they are in parallel, so that can't happen.

In a real circuit, however, this negative feedback effect is overbalanced by a positive feedback effect. The diode with the higher current gets hotter, reducing its forward voltage. This reduced forward voltage forces the lower current diode to also have a lower forward voltage, thus further reducing its current.

Let's see what happens in a real case. We'll do successive approximation. Assume that the two LED packages are in perfect thermal contact. We could assume that they are both hard-mounted to a 1-in. thick piece of copper. (Don't assume that your 10 mile thick aluminum foil has infinite thermal conductivity; it doesn't.) Let's take two LEDs from the same forward voltage bin, which are a maximum of 100 mV different. For ease of reading Figure 5.3, we can assume that one has a forward voltage at 700 mA on the nominal curve of 3.38 V, and the other will then be 3.48 V. (Note the inconsistency in the datasheet; the parameter section says typical forward voltage at 700 mA is 3.60 V.) Now we have 1.4 A to distribute between the two, and they have to have the same forward voltage since they're in parallel. If the nominal device takes 800 mA, it would have a forward voltage of 3.42 V, and the other device gets 600 mA, giving it a forward voltage of 3.32 V + 100 mV = 3.42 V, the same. (We got a little lucky here, guessing the 800:600 ratio right out of the box.)

The nominal device is dissipating 800 mA × 3.42 V = 2.74 W, while the other device is dissipating only 600 mA × 3.42 V = 2.05 W. At 10 °C/W thermal resistance

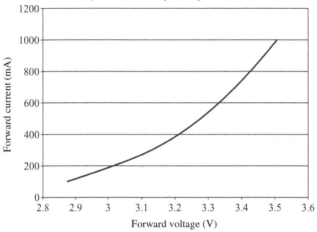

Figure 5.3  Forward voltage as a function of current. (*Source:* "Technical Datasheet DS56, Power Light Source Luxeon Rebel", Philips Lumileds Lighting Company, 2007.)

from junction to case, the die in the one device is up 27 ° from the case, and the other is up only 21°. Now, consider that the forward voltage drops −3 mV/°C. The first device is hotter by 6° than the other, so it has a forward voltage 18 mV lower than the other. This causes a further small increase in the current that the first one is taking, but the change is small enough not to require us to go through a second iteration.

The conclusion is that, at least for this particular LED in this configuration, the positive feedback is not overwhelming but worrisome. One LED gets 200 mA more than the other. It probably won't be damaged, but we're exceeding its 700 mA rating by 14%. But if the metal plate connecting the two were thinner, or they were further away, the temperature difference might start becoming serious. In the worst case where the two LEDs are not connected at all, one could imagine one diode hogging most of the current, getting much hotter than the other, and eventually failing. In the chapter on DC drive circuitry, we'll discuss the feasibility of improving paralleling performance by adding a resistor in series with each LED.

CHAPTER $6$

# PRACTICAL THERMAL MANAGEMENT OF LEDs

IN CHAPTER 5, we discussed the numerous effects that temperature has on LEDs. This leads us to think about how to manage these thermal effects. We start off by giving an easily understood analytical method for calculating thermals. We then turn to an in-depth look at the environment in which LEDs operate, and a variety of methods for keeping them cool.

## INTRODUCTION TO THERMAL ANALYSIS

Think about what happens when you put hot coffee into a ceramic cup. Within 10 min, the coffee is noticeably cooler. If instead you put the coffee into a thermos, it's still hot 10 h later. The unit that measures the difference between the ceramic cup and the thermos is called *thermal resistance*. You can think of it as being analogous to electrical resistance. Instead of resisting the flow of electricity, it stops heat from flowing. The ceramic cup has low thermal resistance, so heat flows through it quickly, and the coffee's heat is lost quickly. The thermos has high thermal resistance, so heat flows through it very slowly and the coffee retains its heat for a long time.

Next think about boiling the water to make the coffee. If you have just a little water, it reaches boiling temperature quickly. If you are making coffee for 10 people on the same burner, it takes much longer to reach boiling temperature. The unit that measures the difference between a little water and a lot of water is called *thermal capacitance*. You can think of it as being analogous to electrical capacitance. Instead of storing up charge and increasing the voltage, it stores up heat and increases the temperature. A little bit of water has a small thermal capacitance, and so the temperature rises quickly. A lot of water has a great thermal capacitance, so its temperature rises slowly.

These examples are perfectly analogous to the principles discussed in this chapter. The underlying reason is that resistors and capacitors are linear devices, and so are thermal resistors and thermal capacitors. The same equations apply to both, only the units are different. We thus have the direct analogy shown in Table 6.1.

*Practical Lighting Design with LEDs*, Second Edition. Ron Lenk and Carol Lenk.
© 2017 by The Institute of Electrical and Electronics Engineers, Inc. Published 2017 by John Wiley & Sons, Inc.

**TABLE 6.1 Analogy between Electrical and Thermal Components**

| Electrical parameter | Electrical unit | Thermal parameter | Thermal unit |
| --- | --- | --- | --- |
| Resistance | Ohm | Thermal resistance | °C/W |
| Capacitance | Farad | Thermal capacitance | J/°C |
| Current | Amp | Thermal power | W |
| Voltage | Volt | Temperature | °C |

You can readily check that this works properly. Ohm's law for electricity says that $V = IR$, which means that volts equal ohms times amps. Similarly, the unit for thermal resistance is °C/W, and the unit for thermal power is W, so their product is temperature with the units of °C.

Furthermore, there is an analog to an RC time constant. Ohms times farads equals seconds. Similarly, thermal resistance times thermal capacitance has units of J/W; since a joule is a watt-second, the product has units of seconds. As an analogy, think of heating up the coffee by heating the end of the spoon you use to stir it. How long it takes for the coffee to heat up to the proper temperature for drinking depends on both the thermal resistance of the spoon (metal has much lower thermal resistance than plastic) and the thermal capacitance of the water (more water has greater thermal capacitance).

## CALCULATION OF THERMAL RESISTANCE

Using this analogy enables us to compute even complex thermal situations. Suppose you have a heat source that can flow through two different paths to the environment. Each path is a thermal resistor, and so they are in parallel. The total heat transfer is the parallel combination of the two thermal resistors, $R = R_1 \times R_2/(R_1 + R_2)$, just like in electricity. The ambient is at a fixed temperature, so it's analogous to a fixed voltage source. No matter how much current (thermal power) you put into it, it's going to stay at the same voltage (temperature). You can thus figure out the temperature at which the heat source will sit. It sits at $V = I \times R + V_{DC}$, and thus $T =$ thermal power × thermal resistance + $T_{ambient}$.

Let's take some specific examples. I have an LED that is dissipating 3 W. Its packaging has a thermal resistance of 10 °C/W from junction to its case. The case has a thermal resistance of 12 °C/W to the ambient temperature, which is at 30 °C. At what temperature is the LED die?

We draw the schematic in Figure 6.1. The ambient temperature is at 30 °C. There's 3 W flowing through the thermal resistance of 12 °C/W, and so the temperature

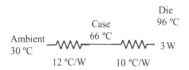

Figure 6.1 Thermal model for LED example.

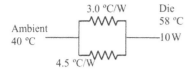

3.0 °C/W   Die
58 °C
Ambient
40 °C ————————————— —10 W

4.5 °C/W

Figure 6.2  Thermal model of two parallel thermal paths.

at the case is $30\,°C + 3\,W \times 12\,°C/W = 66\,°C$. The same 3 W is flowing through the thermal resistance of 10 °C/W, and so the die temperature is at $66\,°C + 3\,W \times 10\,°C/W = 96\,°C$.

Let's try an example with two parallel thermal paths. I have an array of LEDs dissipating 10 W mounted to a heat sink consisting of two flanges. One flange goes to the 40 °C ambient, is 3 cm long, and has a thermal resistivity of 100 °C/(W m). Its thermal resistance is thus $100\,°C/(W\,m) \times 0.03\,m = 3\,°C/W$. The other flange is made of the same material but is 4.5 cm long. Its thermal resistance is thus 4.5 °C/W. What is the LED temperature?

We draw the schematic in Figure 6.2. The two thermal resistances are in parallel from the LED to the ambient, since heat flows through both of them. The ends of both resistances are at the same temperature. Thus, the net thermal resistance $= 3 \times 4.5/(3 + 4.5) = 1.8\,°C/W$. We have 10 W going through this resistance, and so the temperature rise is 18 °C. The ambient is at 40 °C, and so the LEDs are at 58 °C.

Note how the heat flow splits between the two paths. We have a temperature drop of 18 °C. The one path has a thermal resistance of 3.0 °C/W, and so it carries $18\,°C/3.0\,°C/W = 6\,W$. The other path has a thermal resistance of 4.5 °C/W, and so it carries $18\,°C/4.5\,°C/W = 4\,W$. The total heat power is (of course) 10 W, and the lower thermal resistance path carries more of the total heat power.

As a final example, suppose we have an LED in a small light bulb sitting at 25 °C. The LED is dissipating 3 W. We measure its temperature as a function of time using a thermocouple, and get the curve shown in Figure 6.3. What are the thermal resistance and capacitance of the bulb?

The final, steady-state temperature of the LED is 55 °C. This is a 30 °C rise from ambient. Since this is caused by 3 W of power, the thermal resistance is $30\,°C/3\,W = 10\,°C/W$. To compute the thermal capacitance, note that 63% of that 30 °C rise equals 19 °C. The figure shows that it reaches $25\,°C + 19\,°C = 44\,°C$ in 140 s. So the LED rises from ambient to 63% of its final temperature at a time of 140 s, which is its thermal time constant. The thermal capacitance is equal to the thermal time constant

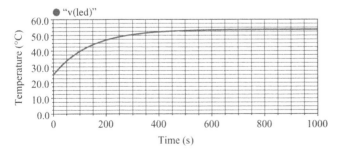

Figure 6.3  LED temperature as a function of time.

divided by the thermal resistance (because $t = R \cdot C$), 140 s/10 °C/W = 14 J/°C. From the thermal time constant, we can also estimate how long we should continue measuring the light bulb to get to a steady-state temperature. In electronics, five time constants is considered a general rule, and so $5 \times 140\,s = 700\,s = 12\,min$ is a reasonable amount of time to measure this system.

## THE AMBIENT

We've been a bit cavalier about the "ambient" temperature. In many cases it's clear enough. Think of setting up a 1 W resistor in your lab to dissipate ½W. Thermally, the ambient temperature is the air in your lab, perhaps 25 °C. If your resistor temperature rises by 50 °C, its temperature will be 75 °C.

But what's really going on here? The resistor has two wires going to your power supply, and those conduct some heat, although not very much. They are also dissipating some heat from their own $I^2R$ designed setup. If the resistor is sitting on a lab bench, then some heat is being conducted by the bench. If the bench is metal, this may be the dominant heat removal mechanism for your resistor. And finally, the air itself carries away heat, but not at an infinite rate. The air too has an equivalent thermal resistance.

We end up with the thermal schematic shown in Figure 6.4. While we don't have values shown for the thermal resistances and capacitances, it's clear that we could measure them. This is quite a complex system, and the "ambient" temperature doesn't have a very clear meaning. So let's simplify this model further. Let's assume that the resistor has a battery built in to it, so it doesn't have any connecting wires to carry or generate heat, and that the resistor is suspended in air by very high thermal resistance strings, so that the bench isn't in contact with it. Now what happens?

Clearly, the temperature rise of the resistor in this case is going to be higher than before, because several of the parallel thermal resistances have been removed. The only one remaining is the air. We know from experience that the resistor surface is not going to sit at 25 °C, the air temperature in the room. It gets hotter. How much hotter depends on a number of factors.

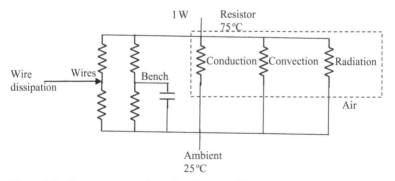

Figure 6.4   There are many thermal paths to ambient.

The ambient air actually has three methods of removing heat from the resistor: conduction, convection, and radiation. The first two depend on there being air; the radiation doesn't. In a practical sense, conduction in air is unimportant for LEDs. The thermal conductivity of air is miniscule, 0.02 W/(m K). This is 1/10 the conductivity of plastics, which aren't very good thermal conductors. In short, air is a pretty good insulator, and we can safely ignore conduction in it.

That leaves convection and radiation. Convection is very complex. It depends on the size and shape of the object in the air, the characteristics of the air, and particularly the size, position, and shape of objects surrounding the object under consideration. Simulations we've seen suggest that the convection flow is usually laminar, not turbulent, which simplifies things a little. But it is too hard to get a practical estimate of convective cooling in air without a specially built simulation model.

Finally, there is radiation. Radiation is governed by the Stefan–Boltzmann law. It says that a black body in free space radiates at a rate set by the fourth power of the temperature. This is somewhat more complicated than it seems at first, because you have to subtract out the effect of the ambient temperature. A body at 298 K (=25 °C) radiates 288 mW/in.$^2$ power, but it also absorbs radiation at the same rate, so that it is in thermal equilibrium. A body at 25 °C sitting in a 25 °C environment doesn't radiate any net power.

## PRACTICAL ESTIMATION OF TEMPERATURE

We can still get an upper bound on the temperature by supposing that there is no convection, and only radiation is removing heat. Once we calculate the temperature based only on radiation, we know that the actual temperature must be lower than this when we add the convection back in. Better yet, there are some empirical reasons to suppose that in the range of interest for LED systems, radiation is about 1/3 the total power removed. The other 2/3 is convection of the air.

Rather than dealing with the math, let's look at Figure 6.5. It gives you an estimate of the temperature of an object dissipating a certain amount of power (W/in.$^2$)

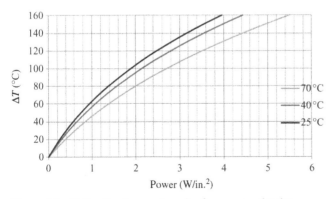

Figure 6.5   Estimating temperature rise from power density.

in air. To use it, you calculate the power the object is dissipating and divide it by the radiating object's surface area. Reading the curve for the appropriate ambient temperature gives you an estimate of the temperature rise of the surface.

Let's take a 5 W resistor as an example. Its surface area is approximately the area of its cylindrical body plus its two end caps. We have area = (8 mm × $\pi$ × 24 mm) + 2 × ($\pi$ × (8 mm/2)$^2$) = 703 mm$^2$ = 1.09 in.$^2$. At 5 W, the power density is 5 W/1.09 in.$^2$ = 4.6 W/in.$^2$. Reading the curve for a 70 °C ambient suggests the resistor body should rise 142 °C above ambient, to 212 °C.

The datasheet shows that the resistor is rated to dissipate 5 W at 70 °C and 0 W at 235 °C (the curve is cut off at 220 °C for unrelated reasons). This means that the 5 W causes a temperature rise of 165 °C. This is reasonably close to our chart's answer.

Let's also try out our original resistor experiment. We have area = (2.8 mm × $\pi$ × 6.35 mm) + 2 × ($\pi$ × (2.8 mm/2)$^2$) = 68 mm$^2$ = 0.11 in.$^2$. At ½ W, the power density is 0.5 W/0.11 in.$^2$ = 4.7 W/in.$^2$. We get the same power density, suggesting the resistor body should rise 142 °C. But of course it doesn't. The resistor is rated for 1 W, so ½ W shouldn't heat it up that much. The problem can be traced to the measurement conditions. For the test, the body of the resistor is elevated above the board. For such a tiny part, the leads contribute significantly to the surface area, and are a significant part of the cooling. The effect wasn't noticeable when we calculated for the 5 W resistor because it is so much bigger.

The conclusion is that for reasonably sized objects dissipating reasonable amounts of power, Figure 6.5 gives a reasonable first guess at surface temperature rise. But there is no substitute for a real measurement.

## HEAT SINKS

Back in Figure 6.4, we showed all of the various thermal paths to ambient. A good way to reduce the temperature of the device is to dramatically lower just one of the resistances. Since they're all in parallel, this one will then dominate the others, just like a 1 kΩ resistor in parallel with a bunch of 100 kΩ resistors is still about 1 kΩ. Since in this case there will be only one significant thermal resistance, this also makes calculation a lot easier.

In this section we will look at reducing the resistance of the thermal conduction path. This is typically done with a heat sink, a piece of metal attached either directly or indirectly to the LEDs. As an example of what can be done, look at the spec for Aavid Thermalloy's part number 569000B00000G (see Fig. 6.6). The big basket in the middle is to allow multiple LEDs to be mounted, the fins are to increase the amount of convective cooling, and the black anodization is to increase the thermal radiation. Together they achieve a thermal impedance of 5.5 °C/W. This will doubtless be the lowest thermal impedance in the system, and so design is easy. If the LEDs are dissipating 8 W, the temperature rise will be 44 °C.

Now of course there's a catch. In order for the heat sink to work as advertised, the LEDs have to make proper contact with it. And in this case, proper contact means proper thermal contact. If the LEDs are loosely suspended above the heat sink, of

Figure 6.6   An LED heat sink. (*Source:* https://www
.aavid.com/products/standard/569000b00000g. Courtesy
Aavid Thermalloy.)

course it won't work properly. You need to attach the LEDs with something that will
both hold them in place and have low thermal resistance to the heat sink.

While there are a number of such methods available, a common one is a thermal
epoxy. A typical one has a thermal conductivity of 1 W/(m K). To figure out the
thermal resistance, *don't* multiply this by the thickness, even though you get the right
units! You need to multiply by the area covered by the epoxy, and *divide* by the
epoxy's thickness. This is thermal conductivity. Then you invert the number to get
thermal resistance.

To take an example, suppose we are bonding a $4\,cm^2$ area, and spread the epoxy
to be 1 mm thick. Then the thermal conductivity will be $1\,W/(m\,K) \times 0.0004\,m^2/$
$0.001\,m = 0.4\,W/K$. The inverse of this is 2.5 K/W. This is in series with the thermal
resistance of the heat sink, giving a total thermal resistance of 8.0 K/W from LED case
to ambient. With our 8 W example above, the temperature rise is actually 64 °C, rather
than 44 °C.

Note that increasing the area of the thermal contact increases the conductivity,
which decreases thermal resistance, just as you would expect. And similarly,
decreasing the thickness of the epoxy also decreases thermal resistance. So the
goal should be to get the maximum surface contact area possible, and then spread the
epoxy as thin as possible.

While we're on the subject of heat sinks, we should mention metal-core PCBs
(MCPCBs). These are made by specialist manufacturers, and consist of a regular PCB
bonded to a piece of metal, usually aluminum. The bonding is usually isolated.
MCPCBs have found some favor for mounting LEDs. The attraction of the method is
that thermal performance is better than that sometimes accomplished by "do-it-
yourself" methods. The downside, of course, is that the cost is higher than that of
regular PCB. The authors have considered using them in a number of projects, but
usually end up not doing so because of the additional cost.

## FANS

Another thermal resistance that can be minimized is the convection. Convection in air
is moderately effective at cooling, but it can be dramatically enhanced by a fan (for our
practical purposes, a blower is the same as a fan). Fans and blowers work by forcing
air to move across the hot surface. But how do you select a fan?

Fans are typically rated by how much air they move, with units of liters/minute. The reality is that there are a lot of complex motions of the air that determine how much cooling you get. The size and shape of the fan, the size and shape of the object being cooled, and its orientation to the fan, as well as other objects in the path of the air flow, all contribute to the actual cooling. So the best we can offer is an estimate of how much fan you need.

From an article by Mike Turner (Turner 1996), we can estimate the flow rate from the equation $G = P/(\rho C_p \Delta T)$, with $G$ the volumetric flow rate in $m^3$/s, $\rho$ and $C_p$ characteristics of air, and $\Delta T$ the temperature rise. Plugging in air values at 25 °C, and converting to liters and minutes, we end up with the estimate

$$G(\text{l/min}) = \frac{12P(\text{W})}{\Delta T(\text{K})}.$$

Note that this is independent of the surface area of the object being cooled, and so we are assuming that all of the air blows past it. As an example, consider the 5 W resistor mentioned earlier. It rose 165 °C in 70 °C still air. Suppose we want the temperature rise to be a much milder 30 °C. Then, we need airflow of $12 \times 5$ W/30 K = 2 l/min. Looking at actual devices, we see the Sunon[1] UF3A3-500 has a rated flow of 3.43 l/min, and so will probably do just fine. Of course, this formula is only an estimate; you need to actually build the system to verify that temperatures are what you calculate.

Now there are a number of issues with fans. In the first place, the air has to have some place to go. If you put a fan into a totally enclosed system, it will help to make sure the whole inside is at the same temperature, but it won't help much with getting the heat out to the ambient. To work best, you need to have an inlet and an outlet for the air. This may or may not be an aesthetic system design problem.

For one thing, you have a high-speed blade whirling around (17,000 rpm for the fan we just picked). What happens if a piece of dust or paper gets sucked in? Aside from affecting your cooling, it could also cause the fan to stall. Is the fan protected so that it doesn't burn up when it stalls indefinitely? PTCs are often used for this purpose. And what about safety issues? What if a child sticks his finger into the fan? This must be considered when using a fan as a coolant.

The fan also requires power. AC can be convenient for powering fans, but then you have to make sure customers can't come into contact with the live AC. It's much more common to use DC brushless fans, which run on 5 or 12 V and also have reduced EMI compared with other types. But the 5 V has to come from somewhere, this somewhere presumably being the LEDs' ballast. Do you have 5 or 12 V already in the ballast, and if not, how are you going to generate it? Fans can draw substantial current; the one we just picked needs 100 mA at 3 V, 300 mW. This is more than you want to generate from a zener. And now you've added an extra 300 mW to the power dissipation of your light. That means additional cooling is needed, and that there's an additional drop in efficacy of the light.

---

[1] http://www.sunon.com/mmfan/mmfan_e.htm

A final concern is acoustic noise from the fan. Everyone is familiar with noisy fans in computers. Less noisy ones cost more. How will your customers react to a light that has a fan running all the time?

# RADIATION ENHANCEMENT

One more area that you should consider for reduction of thermal resistance is the thermal radiation. You can't affect the laws of physics, but notice that the estimated temperature rise in Figure 6.5 depends on both the amount of power and the surface area of the object. If you could increase the surface area, that would decrease the temperature.

To get a feel for how much benefit this could have, consider that 5 W resistor again. Its nominal size is 8 mm diameter by 24 mm length. But the maximum size allowed by the tolerance is 9 mm diameter by 25.5 mm length. The nominal values have a surface area of 703 mm$^2$, while the maximum is 848 mm$^2$. This is an increase of 21%. It drops us from 4.6 W/in.$^2$ to 3.8 W/in.$^2$, dropping our expected temperature rise from 165 to 124 °C.

The conclusion here is that even small increases in size can have dramatic effects on temperature. The manufacturer of the resistor could increase the power rating of his device without affecting its size by holding the mechanical tolerances tighter, and specifying the nominal dimensions more toward the top of the current range. It's worth asking whether your device could be 10% bigger without customer rejection.

# REMOVING HEAT FROM THE DRIVE CIRCUITRY

While the focus on this chapter has been on thermal management of the LEDs, you may also need to do some thermal management of the ballast. Consider a ballast delivering 30 W to a set of LEDs. Getting rid of the heat from the LEDs is probably the most important concern you have, but it shouldn't be the only one. If the ballast is 85% efficient, the power it dissipates is (30 W/0.85) − 30 W = 5.3 W, 18% of the power the LEDs are getting.

What are the alternatives? You might be told to increase the efficiency. If the efficiency were 100%, then it wouldn't dissipate any power at all! Of course, 100% efficiency is physically impossible. And as we'll see in subsequent chapters, it is hard to increase efficiency very much. A good compromise would be 85%. You can get to 90%, or maybe even 92–93%, but it costs increasingly more money and design time to do so. At some point, it doesn't make sense to pursue efficiency improvement any further.

There are specific actions you can take to remove heat from the ballast. The first is simply to realize that only a few components in the ballast generate significant heat. These are the power components, typically the transistor and diode, though the IC may in some cases also dissipate significant power. Putting a good amount of copper on the PCB, including a ground plane, can help to distribute the heat more evenly throughout the ballast, helping the temperature of these components.

Another important thing to realize is that ballasts are fairly immune to high temperatures. Most components used in a ballast are rated to 125 °C. The important exception is electrolytic capacitors. These can have lifetimes very seriously degraded by elevated temperature, and can become the lifetime-determining component in the entire light. Methods of dealing with this, as well as ways of not using electrolytic capacitors, are detailed in subsequent chapters. Given this, there may not be a grave disadvantage to letting the ballast run hot, as long as customers can't burn themselves.

The same things that can be done to alleviate temperature in an LED system can also be applied to the ballast. The power components can have heat sinks attached to them. The ballast can have a heat sink attached to it on the outside case. In extreme cases, a fan can be used to blow air over the ballast case. But again, the easiest thing to do is to make the ballast mechanically larger. This has two positive effects. It allows the heat-generating power components to be placed further away from the heat-sensitive components, and it provides increased surface area for the ambient to cool the ballast.

CHAPTER 7

# PRACTICAL DC DRIVE CIRCUITRY FOR LEDs

**L**EDS NEED to be electrically driven in order to emit light. In this and the next chapter, we're going to discuss how to design drive circuitry for LEDs. In this chapter we're specifically interested in DC drive circuitry. A typical DC source is a battery, as, for example, used in a flashlight. It could also be the output of a switching converter, for example, 12VDC. In any case, what distinguishes DC from AC for practical purposes is that DC typically has a much lower voltage than AC. This means that the regulations governing usage are much easier to comply with. There's no EMI to worry about, the voltages are generally "safe" according to UL and the input voltage is generally very steady. On the downside, the currents in a DC drive are higher, and the source impedance becomes an important factor in the design.

## BASIC IDEAS

The fundamental determinant of what type of converter to use for a DC drive is set by the relative values of the supply voltage and the LED voltage. There are three cases:

1. The supply voltage is always higher than the LED voltage.
2. The supply voltage is always lower than the LED voltage.
3. Sometimes the supply voltage is higher than the LED voltage, and sometimes it's lower.

When we use the word "always," this includes variation of the supply voltage with time, as well as variation of the forward voltage of the LEDs due to temperature, binning, and so on. For example, the supply voltage is always higher than the LED voltage only if there is *never* a time in operation when it is lower.

In the first case, the supply voltage is always higher than the LED voltage. A typical example of this would be a flashlight using four D cells, and running a single 3 W white LED. The D cells are 1.5 V when new, and something like 0.9 V when at

*Practical Lighting Design with LEDs*, Second Edition. Ron Lenk and Carol Lenk.
© 2017 by The Institute of Electrical and Electronics Engineers, Inc. Published 2017 by John Wiley & Sons, Inc.

end of life, giving a battery voltage range of 3.6–6.0 V. The LED's forward voltage is 3.2 V at 350 mA, and decreases with increasing temperature. In cases like this, a buck converter would normally be used.

In the second case, the supply voltage is always lower than the LED voltage. A typical example of this would be a flashlight using two C cells, driving a set of seven 5 mm white LEDs. The C cells have the same voltage range as Ds, so the battery voltage range in this case is 1.8–3.0 V. The LEDs have a voltage of 3.2 V each, for a total string voltage of 22.4 V. In cases like this, a boost converter would normally be used.

Finally, in the third case, the supply voltage is sometimes higher and sometimes lower than the LED voltage. An example of this would be the same as the example for number one, but using three "D" cells rather than four. Now the battery voltage range is 2.7–4.5 V. The LED voltage is right in the middle of this range. For a case like this, a buck-boost converter would typically be used.

## BATTERY BASICS

All three of our examples used batteries. Therefore, we need some basic information about batteries. First let's examine terminology. In common speech, we use the word "battery" to describe both a D-size battery that you buy in the supermarket and the big battery that powers a laptop computer. Technically, however, the packages the supermarket is selling contain cells, not batteries. A battery is the end device that powers something. It often consists of two or more cells, usually in series. But sometimes, as in children's toys, it consists of a single cell, hence the confusion. Thus, a laptop really does have a battery. It has three or four 4.3 V lithium cells in series. Your flashlight with two D cells also is using a battery. We'll try to consistently maintain the difference between the words.

The most basic piece of electrical information about a cell or battery is that it is a voltage source. To zero order approximation, an alkaline cell of any size (D, C, AA, AAA, etc.) produces 1.2 V. To the same approximation, a lithium cell produces 3.6 V. The reality is considerably more complicated. Cells do in fact change their output voltage, depending on load current, temperature, state of charge, and age. And different types of cells have different characteristics as well. Some understanding of all these aspects is necessary to be able to properly design a DC drive for LEDs.

Cells are actually complicated electrochemical systems. The "chemical" part of this word is referring to how the cell stores energy. Cells work differently from capacitors. Capacitors are purely electrical. They store energy in an electric field. Typical energy levels for a capacitor can be described by the image of a 1 µF capacitor charged to 5 V. The energy it stores is

$$E = \frac{1}{2}CV^2 = \frac{1}{2}1\,\mu\text{F}\,(5\,\text{V})^2 = 12.5\,\mu\text{J}.$$

Cells store energy in chemicals. Typical energy for a cell can be seen by thinking of a 1.2 V cell with 200 mA h of charge. (Milliamp-hour is the usual rating for cell capacity.) The energy stored is approximately

$$E \approx V \cdot Q = 1.2\,\text{V} \cdot 200\,\text{mA h} \cdot 3600\,\text{s/h} = 864\,\text{J}.$$

The cell stores some eight orders of magnitude more energy than the capacitor. This is why batteries are used rather than capacitors to run LEDs for lighting or for any device requiring significant power.

While we don't need to know exactly how the chemistry in the cell works, it's important to know that there are a couple of different kinds on the market. We've already mentioned the two most important, alkaline and lithium (nickel metal hydride (NiMH) has been largely superseded by lithium). The C and D cells sold in the store are typically alkaline. These are usually single-usage types. Their fully charged voltage is 1.5 V, and end-of-life voltage is about 900 mV. You can actually run them lower than this, but there is very little energy left in them below 900 mV.

The other common type of cell is a lithium cell. These are almost always rechargeable. Their fully charged voltage is about 4.3 V, and discharged voltage is usually considered to be about 2.7–3.0 V. Lithium cells are considerably more sensitive than alkaline cells. For example, they might explode if you continue to charge them after they're fully charged. And they can be permanently damaged if you discharge them below about 2.4 V. The bottom line here is that practical lithium batteries almost always use a special IC to control them, to take care of all the problems they can present. For our purposes, we'll simply assume that this circuitry is already in place, and that the LED drive we're designing works with the output of this circuitry.

Now we can examine the performance of batteries. As we've already noted, the output voltage of a cell is constant only to zero-order approximation. What really happens is that for small loads, the voltage is close to a constant. But as the load becomes bigger, the voltage drops, approximately linearly for medium loads, as though there were a resistor in series. When you get to heavy loads, the voltage drop is large and nonlinear, something like a hyperbolic tangent (see Fig. 7.1). In a practical sense, batteries for LED lighting are almost never run at heavy loads. A little thought shows why. If you run the battery very hard, it will discharge quickly. For most lighting applications, you want the light to be on for a long time. Therefore, you almost always run the battery in its light or medium load ranges. And, thus, estimating the battery voltage as an ideal battery plus a resistor is reasonable for most of our applications.

But what is a light or medium load? And what resistor value should you assume? There's no standard answer. It varies from type to type and to some extent,

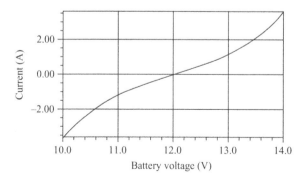

Figure 7.1 *I–V* curve of 12 V battery. (*Source:* Lenk (1998).)

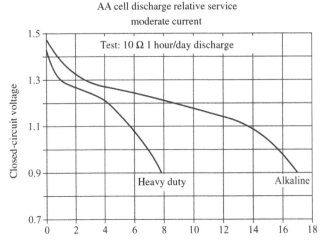

Figure 7.2 Alkaline cell battery voltage as a function of time with a resistive load. (*Source: Rayovac, OEM 151 (R-3/99), "Application Notes & Product Data Sheet," "Primary Batteries—Alkaline & Heavy Duty," Figure 1. Property of Spectrum Brands, Inc.*)

even from manufacturer to manufacturer for the same type. Let's look at an example to see how to figure it out. Looking at Figure 7.2, the manufacturer supposes a constant $10\,\Omega$ to constitute a moderate load for an AA alkaline, a 2.2 Ahr cell (we're looking at the "alkaline" cell curve in Figure 7.2, not the "heavy duty" cell curve). Much of the time, the cell sits at 1.2 V, and so the current for a typical moderate load is around 120 mA for this cell. Note that as the cell discharges, the output voltage drops.

Now we can estimate the approximate series resistance in this cell. The datasheet says that the fresh voltage of the cell is 1.55 V. Figure 7.2 shows the $10\,\Omega$ load at the beginning of discharge seeing a cell voltage of 1.47 V. This is a drop inside the cell of 80 mV in response to a current of 120 mA. We thus estimate the internal cell resistance as $80\,mV/120\,mA = 667\,m\Omega$. It's unfortunate we can't make the same estimate for the cell resistance when it has almost finished discharging, but we'll assume that it's not radically different.

Another factor affecting batteries is ambient temperature. As cells get colder, their voltage drops and their impedance rises. This particular alkaline is rated from $-30$ to $+55\,°C$. Below $-30\,°C$, the chemicals inside the cell start to freeze. Practically, what this means is that the impedance becomes so high that the cell can't source any current. And what happens above $55\,°C$? Unless you're considerably above $55\,°C$, probably nothing catastrophic will happen. But the self-discharge rate becomes so high that it seriously undermines operating life from the cell. (Batteries sitting on the shelf do discharge themselves gradually.)

Finally, there are aging effects on rechargeable cells. As you charge and discharge them repeatedly, they gradually lose their ability to store charge. After some hundreds of cycles, they will lose so much capacity that they are considered dead. In the old days, nickel-metal hydride batteries also had a memory effect. If they were only partially discharged before being recharged, they lost some of their storage capacity. Fortunately, lithium cells don't have this sort of problem.

# OVERVIEW OF SMPS

Switch-mode power supplies (SMPS) are used almost universally to convert a source of power into a form suitable for a load. Common ones include the "wall-wart" that plugs into the wall to convert 120 or 240 VAC into 5 VDC to charge your cell phone, and the one on the PCB of your computer that converts the 3.3 VDC from the silver box to 1 V to run the processor. Other converters deliver a constant current rather than a constant voltage, and this is the kind we will be designing for LED lighting. The fundamental distinction between all types of SMPS is whether the power source is AC or DC, and we will be following this distinction by treating the two in separate chapters. This chapter is about DC.

SMPS operate in switch mode, hence the name (see Lenk, 1998). What this means specifically is that there are one or more transistors in the circuit, which switch on or off to control the operation of (usually) an inductor. Transistors *could* be operated in linear mode instead of switch mode, but this is inefficient. Referring to Figure 7.3, we see the problem. The transistor is basically acting as a linear regulator. The input voltage of 12 V is dropped to 5 V across the transistor, which therefore has 7 V across it. If the output current is 1 A, then the power dissipation in the transistor is $7\,\mathrm{V}\cdot 1\,\mathrm{A}=7\,\mathrm{W}$, and the efficiency is $\eta=P_{out}/P_{in}=(5\,\mathrm{V}\cdot 1\,\mathrm{A})/(12\,\mathrm{V}\cdot 1\,\mathrm{A})=42\%$.

In switch mode, the circuitry is much more efficient. Referring now to Figure 7.4, the transistor is shown as a switch, which is either open or shorted. The efficiency of the SMPS comes from operating the transistor only in these two modes, and trying to get from one to the other as fast as possible. When the switch is shorted (the transistor is on), as in part (a), the input voltage is directly applied to the inductor. The diode is reverse-biased and so is off. Since the voltage on the switch side of the inductor is higher than the output voltage, the inductor current increases (remember that $V=L\,\mathrm{d}I/\mathrm{d}t$).

In the next step, shown in part (b), the switch is open (the transistor is off). The input voltage is disconnected from the rest of the circuitry. Now the voltage across the inductor is the other way. The output voltage is higher on the output than on the input of the inductor, so the current decreases. The current continues to flow in the inductor, and since it has to come from somewhere, it comes through the diode. The voltage on that node drops until the diode can conduct. The diode becomes forward-biased and turns on. That means the voltage on that side of the inductor drops to a diode drop below ground, about −0.7 V.

Now here's the crucial step. The inductor current is thus forced to alternately increase and decrease each cycle. The inductor current goes into a capacitor, which is alternately slightly charged and discharged each cycle. By controlling how long the inductor current increases versus how long it decreases, the average current can be set. And it is set in such a way as to produce the desired average voltage on the output capacitor. The setting is done by a feedback circuit, the PWM (pulse-width

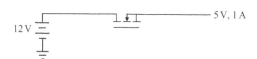

Figure 7.3   Operating a transistor in linear mode is inefficient.

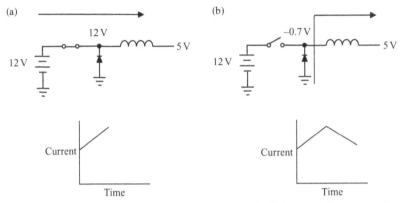

Figure 7.4    When the Transistor ($t$) is on, current in the inductor ($I$) increases; when the transistor is off, current in the inductor decreases.

modulation) controller, which in practice is always an IC. The controller measures the output voltage, and adjusts the amount of time the switch is on to keep the output voltage constant. The ratio of the on-time of the switch to the total time (on-time plus off-time) is called the duty cycle (DC, although this has to be distinguished from "direct current" by context).

There are quite a number of different ways of hooking up the transistor, inductor, diode, and capacitor, including methods that use multiple devices. All of the different configurations are generically called SMPS topologies. There are also a variety of modulation schemes, such as the constant-frequency control we've been discussing, as well as constant on-time control and others. But all of these systems work on the same basis, modulating the current in an inductor to regulate the output. In the next few sections, we will cover some of the basics of the topologies that will be suitable for driving LEDs from DC sources. We'll cover these designs in much more detail in Chapters 10–13.

# BUCK

Let's cover the first case, where the source voltage is always higher than the LED voltage. The topology we will use is the buck converter, which is what we have been discussing in the previous section. Recall how this worked. To make the inductor current increase, we needed to apply a voltage to it that was higher than the output voltage. If the voltage applied was lower than the output voltage, the current would decrease, not increase, and the output voltage would drop, no matter how long the transistor was on. Eventually, the output voltage would equal the input voltage.

Thus, a buck converter can only convert an input voltage to a lower output voltage. You can't get a higher voltage out of this topology than you put in. This is exactly the way the first case works. As an example circuit, we're going to look at National Semiconductor LM3405. This part is convenient to use because the switch is integrated inside the IC, and the whole thing fits in a 6-pin SOT, a tiny package. We

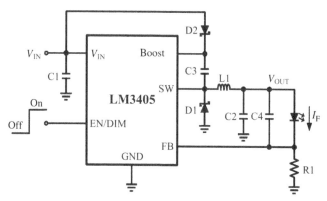

Figure 7.5   LM3405 schematic for buck. (*Source:* LM3405 datasheet, National Semiconductor, February 2007.)

will be using this circuit in a flashlight design in Chapter 11, converting the USB
output of a computer to drive a single power LED.

Look at Figure 7.5, taken from the front page of the LM3405 datasheet. The
input voltage can be as high as 20 V on this device, and so the IC is powered directly
from the input on the $V_{IN}$ pin. The transistor for the buck is inside the IC, between the
$V_{IN}$ and the SW (switch) pins. The LM3405 switches the transistor on and off at a
constant frequency of about 1.6 MHz. This connects and disconnects the SW pin to
the input voltage. When the transistor is on, the voltage is positive across L1, and
when the transistor is off, the voltage on SW swings down to a diode drop below
ground, turning on D1. The current in L1 thus increases and decreases. Note that the
high switching frequency of the LM3405 means the value of L1 can be very small,
typically around 10 μH.

The current in L1 is smoothed by C2, and is fed to the LED. C4 is basically in
parallel with C2, but also has some function for stability of the circuit. The current is
sensed by R1, which produces a voltage proportional to the current and is fed back to
the FB (feedback) pin. The IC controls the duty cycle of its internal switch to produce
a voltage of 205 mV on the FB pin. This voltage is intentionally low in order to avoid
significant power loss in R1 (1 A · 205 mV = 205 mW, which can be handled by a
¼W resistor). This may possibly require an RC filter between R1 and the FB pin, to
reduce noise fed back to the IC. Other ICs have even lower feedback voltages. You
should generally avoid using a normal IC used for producing a fixed output voltage.
For example, a 2.5 V feedback at 1 A would lose 2.5 W in the resistor.

The LM3405 can be controlled on and off with a digital signal on the EN/DIM
pin (1.8 V on, 0.4 V off). The final elements in Figure 7.5 are D2 and C3. These are
used as a charge pump to produce a voltage on the BOOST pin that is higher than the
input voltage. What is this used for? Remember that there is a MOSFET inside the IC
between $V_{IN}$ and SW. When the MOSFET is turned on, we want the voltage on SW to
be as close as possible to the input voltage, as otherwise there is a voltage drop across
the transistor, leading to power loss. But to turn on an N-channel MOSFET, the gate
voltage has to be higher than the source voltage. The BOOST pin provides that higher
voltage.

## BOOST

In the second case, the input voltage is always lower than the output voltage. The topology we will use is called a boost. As the name suggests, it takes the input voltage and boosts it up to a higher voltage. And just as the buck is incapable of producing a higher voltage output than its input, the boost is unable to produce an output that is (significantly) lower than its input. As an example circuit, we're going to use the Fairchild FAN5333. This part is convenient to use because the switch is again integrated inside the IC, and the whole thing fits in a 5-pin SOT. We will be using this circuit in a flashlight design in Chapter 10, boosting the output of 2 'D' alkaline cells to drive a 3 W LED.

To understand how a boost works, look at Figure 7.6. Once again, the input voltage goes to the $V_{IN}$ pin, this time limited by the manufacturer to 6 V. Now in this case, the transistor, diode, and inductor are still present, but are connected together differently. Here's how this works. The input voltage is always attached to the inductor. The reason the inductor current doesn't continue to increase is because the voltage on its other side is higher than the input voltage (that's why the boost has to have higher output than input voltage). The transistor is operated by periodically pulling the SW pin to ground. When the transistor is on, SW is at zero volts, and so the inductor current increases. When the transistor is off, SW goes up to a diode drop above the output voltage, and so the inductor current decreases.

The purpose of the diode D1 is thus clear. When the switch is on, the SW pin is pulled to ground. If the diode wasn't there, that would pull the output to ground as well. The current from the inductor goes through D1 and into C1 and the load. The current is sensed by R1, producing a feedback signal at the FB pin. The FAN5333A controls the duty cycle of its internal transistor to produce a regulated voltage at the FB pin (110 mV), corresponding to the desired current through the LED. You should note that when the switch is on, the inductor current is going to ground. To keep the LED on during this time, there has to be enough energy stored in capacitor C1 to ensure that the voltage across the LED doesn't change significantly.

In this particular schematic, the shutdown pin, SHDN, is connected high, so that the IC is always on. The only other aspect to note in this design is that no charge pump

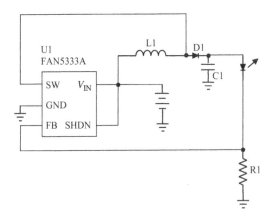

Figure 7.6 FAN5333A schematic for boost. (*Source:* FAN5333A datasheet, Fairchild Semiconductor, August 2005.)

circuit is required. The internal MOSFET goes from SW to ground, and so the input voltage is high enough to run its gate.

## BUCK-BOOST

We've saved the best for last. The buck converter can only be used when the output voltage is less than the input voltage. Conversely, the boost converter can only be used when the output voltage is greater than the input. What do you use when the input is sometimes higher than, and at other times lower than, the output? This situation arises because battery voltages are dependent on a number of factors, including their state of charge and temperature. So a fresh battery may have a voltage greater than the LED it's driving, but when it's been mostly discharged, its voltage may be lower.

The traditional circuit for this power supply is called a buck-boost. As the name suggests, it is a buck and a boost stuck together. This involves two transistors and two diodes, one set for each. Conceptually, it works by turning off the buck and running the boost when the input voltage is low, and by turning off the boost and running the buck when the input is high.

For many LED drivers, however, this may be too expensive. Four external power devices eat up quite a bit of board space and money; using integrated switches costs even more. In this section, we're instead going to do something more clever. We will use the Supertex HV9910 to create a circuit that works regardless of the magnitude of the input or output voltages. We will be using this circuit in a flashlight design in Chapter 10, converting the output of a car battery to run an LED taillight.

To understand how this works, look at Figure 7.7. Once again, the input voltage goes to the $V_{IN}$ pin, although this time it can be as high as 450 V (the HV9910 is really designed for AC off-line application). Now in this case, the transistor, diode, and

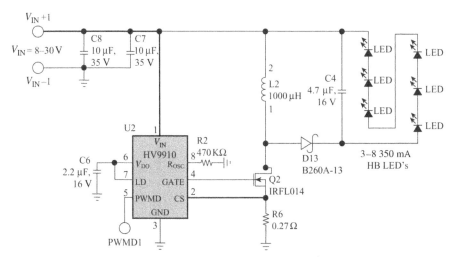

Figure 7.7    HV9910 schematic for buck-boost. (*Source:* HV9910 datasheet, Supertex Inc., 2006.)

inductor are connected together exactly the same as they are in the boost converter. The transistor is external, but this has no bearing on the operation. What's different about this circuit from both the LM3405 and the FAN5333A circuits is the position of the load, the LEDs. In the previous circuits, the LEDs were connected from the output to the ground. In this circuit, by contrast, they are connected from the output *back to the input*. How does this work?

LEDs are current devices. As long as they have current flowing from their anode to cathode, they light up. It does not matter to their operation where ground is. So the idea of this circuit is to use a boost converter to produce a voltage higher than the input, and then run the LEDs from this. Since the output voltage is higher than the input, the LEDs can be referenced to the input rather than ground. The big advantage of this circuit is that it doesn't matter what the voltage across the LEDs is. It can be lower in magnitude than the input voltage is to ground, or higher. Since the top of the LED string is at higher voltage than the input, the boost operates properly no matter what.

To complete the circuit, capacitor C4 is in parallel with the diodes. It too is not ground referenced. Since this HV9910 circuit is indeed a boost, during Q2's on time the inductor current goes to ground, not through the LEDs. Thus, C4 is there to provide current during this off-time.

Now while this circuit has notable advantages, it also has some downsides, which are fortunately taken care of by the IC. In the first place, the FAN5333A and the LM3405 measure the LED current by measuring the voltage across a ground-referenced resistor. Since the IC is also ground-referenced, this is convenient. The IC can just put the measured voltage directly into a comparator. But for this HV9910 circuit, the LEDs are above ground, and so sticking a resistor in series with them would require accurately measuring the difference between two high voltages, which is difficult. The HV9910 instead measures the LED current by changing the way it controls the switch.

Both the FAN5333A and the LM3405 are standard SMPS (technically, they use current-mode control, but that doesn't matter to us here). They compare how much current there is with how much there is supposed to be. If these two are different, they send the difference to an amplifier that adjusts the duty cycle of the switch. If the current is the right amount, the duty cycle stays the same. The way, the duty cycle is adjusted to give exactly the right amount of current to the LEDs.

The HV9910 works differently (it uses peak current control). It turns on the switch and then measures the current as it is increasing through the inductor. When it reaches a certain level, it turns off the switch. The average current is thus less than the set point; the value of the inductance determines just how much less. The point is that this still regulates the LED current, just in a different way than the other two ICs. And this difference is what allows the HV9910 to work in this topology. Since it just looks for a peak current value, it can measure the current through the (external) switch. The inductor current ramps up when the switch is on, and it goes through the switch and also through a current sense resistor. The current sense resistor is referenced to ground, and so the control signal is once again straightforward. When the current through the inductor reaches the preset maximum level, the IC turns the switch off. Then the current flows through the LEDs. When the current goes through the LEDs, it decreases, and we have actually measured the maximum current. Thus, we are again regulating something related to the LED current, and the difference between this value and the real current is controlled by the inductance value.

The other drawback of this circuit configuration is that the voltage that is seen by the transistor and diode is higher than the input voltage. In fact, they see the input voltage plus the LED string voltage. But unless the string is very long, this is not very much voltage and so parts are readily available to handle it at no extra cost or space. The high voltage also means somewhat higher losses in these two components, but again this may not be practically that significant in real applications.

# INPUT VOLTAGE LIMIT

For the three ICs used in this chapter, we've mentioned the input voltage, that is, the maximum voltage that can be applied to the $V_{IN}$ pin. But how do you practically decide what voltage IC you need?

The first thing to remember is that the maximum input voltage for the IC is not necessarily the same as the maximum input voltage the circuit can handle. Indeed, they are usually different. Although our selection of the FAN5333A and the LM3405 might suggest otherwise, most controller ICs do not have integrated power devices. In this case, then, the external power devices can be very high voltage, while the IC runs off a low supply voltage. The reason to use this method is that the high-voltage IC process is expensive. It is much cheaper to make the IC low voltage. In this case, then, the choice of IC input voltage is determined by where the supply voltage comes from. If the DC input voltage is low, say below 40 V, then you can use an IC whose input will directly tolerate that voltage. If the DC input voltage is higher, then you probably need to use a low-voltage IC, and generate the supply voltage inside the circuit. In this case, a 12 V IC might be a good choice, 12 V being a common, inexpensive process. (Why not use a 5 V IC? Because 5 V is not always enough to run the gate of the MOSFET, and so the IC plus the transistor cost may be higher, even though the IC by itself is cheaper.)

To generate the supply voltage for the IC, should this be necessary, the easiest thing to do is just put a resistor in series with a zener (don't forget a bypass capacitor on the $V_{IN}$ pin to counter noise). For modern ICs drawing low $V_{IN}$ current, this won't dissipate too much power in the resistor. If for some reason you need more than a couple of milliamps supply current, then you probably have to do something more complicated, such as adding a winding to a transformer to make an internal power supply.

It is only when the switch is integrated inside the IC that the IC needs to be able to handle the full input voltage. But even then, economics determines that the $V_{IN}$ pin will usually want a low voltage, and the high voltage is reserved just for the pins with the power devices attached to them. Thus, for example, the FAN5333A has a 35 V maximum on its SW pin, but 6 V maximum on $V_{IN}$. The LM3405 has both $V_{IN}$ and SW rated at 20 V, but this is because the buck transistor is attached to $V_{IN}$. The HV9910 is the exception in that it has its $V_{IN}$ pin rated at 450 V. But even this then regulates it down to 7.5 V on its $V_{DD}$ pin, which is used for the internal circuitry. The high input voltage rating on the $V_{IN}$ pin was an intentional choice to avoid needing the external resistor and zener, and to avoid certain start-up issues. Otherwise, it is unnecessary, as the transistor is external to the IC.

## DIMMING

In Chapter 2 we've already mentioned the two popular ways of dimming LED circuits, PWMing and analog. We can now look at these methods more closely, and see how to apply them to our circuits.

Because the optical characteristics of the LEDs are (at present) dependent on drive current, it is desirable to operate them at their rated current. The way to dim LEDs while holding them at a constant current is to turn them on and off periodically, as shown schematically in Figure 7.8. This is called pulse width modulation (PWM), the same as for controlling ICs. During their on-time, the LEDs are run at their rated current. The percentage of time they are on is called their duty cycle (abbreviated DC, and not to be confused with the power source type). Since the rest of the time during the cycle (or period) they are off, their *average* current is lower. In fact, the average current is their on-time current multiplied by the duty cycle.

Now to avoid flickering of the LED, the frequency of the PWM has to be above about 60 Hz. The ICs we've used in this chapter easily accomplish this. For example, the LM3405 can be PWMed up to about 5 kHz. Since it's running at 1.6 MHz, you might suppose that it could be modulated much faster. But what is actually happening is that the IC is being turned on and off, and it takes about 100 μs to turn on. This limits the PWM frequency, because full LED current won't be present until the IC is fully on. The same thing limits the FAN5333A to only about 1 kHz. If you use ICs that were not designed with this type of modulation in mind, it can take so long for the IC to turn on that the modulation frequency will have to be in the visible range. This sort of IC is then unsuitable for dimming LEDs.

The other generally applicable method of dimming LEDs is to reduce their drive current. Many applications will not be affected by color shifts induced by reduced drive current. As shown in Figure 7.9, a resistor divider from the current sense resistor to the feedback of the IC is all this method takes, although you may want a capacitor to control noise. Bear in mind how this circuit works. As the variable resistor R3 is decreased in value, the voltage presented to the IC is lessened. This makes it generate more current through the LEDs. So in this case, the minimum LED current is set by R3, with the divider set at maximum. As the divider goes to zero, the LED current will try to go to infinity, and so R4 is added in series with the divider to provide a maximum for the current.

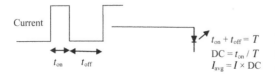

$t_{on} + t_{off} = T$
$DC = t_{on} / T$
$I_{avg} = I \times DC$

Figure 7.8  Pulse width modulation turns the current rapidly on and off to get an average current.

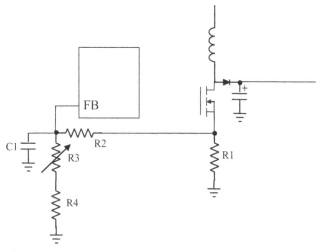

Figure 7.9  Dimming circuit.

# BALLAST LIFETIME

In Chapter 5 we talked about LED lifetime and how it depends on temperature and other factors. But in a lighting system, the lifetime of the entire system needs to be considered. Assuming that the LED lifetime is adequately under control, the drive circuit may actually determine the lifetime of the system. Formally, the lifetimes of the individual parts add like paralleled resistors. Thus, if one part has a much lower lifetime than the other, the system's lifetime is controlled by that part.

$$\text{System lifetime} = \frac{1}{(1/\text{Part \#1 lifetime}) + (1/\text{Part \#2 lifetime})}$$

The lifetime of a properly designed drive circuit is usually very long. ICs in particular can last decades if they are not subjected to excessive voltages. And the power components, MOSFETs and diodes, are equally robust under the same conditions (although this is different in AC drives, see the next chapter). The real limiting factor on the driver lifetime is the input capacitor.

In real DC drive circuits, the input voltage is always buffered by a capacitor. The FAN5333A circuit in this chapter doesn't show it, but it is necessary. The reason, as outlined in the battery section of this chapter, is the fairly high impedance of batteries at typical switching frequency. *How much* capacitance is needed depends on the power level of the driver and the switching frequency. You don't want the input voltage to drop very much during the time when the switching transistor is on. So higher input current means you need more capacitance. Conversely, switching at a higher frequency means that the current is drawn for less time, and so you need less capacitance.

The problem for lifetime is the electrolytic capacitors. The factors to watch out for are the rated lifetime of the electrolytic capacitor, its rated temperature, rated voltage, and rated ripple current. For example, the HV9910 circuit shows two 10 µF capacitors at the input. Let's suppose we use 35 V electrolytic capacitors rated at 1000 h at 85 °C. These are probably the cheapest ones we can get. Now if the drive circuit is encased so that the air temperature runs at 65 °C, then the capacitor is being run 20 °C below its rating. For each 10 °C drop below rating, the capacitor lifetime approximately doubles. So here, the lifetime of each capacitor is about 1000 h × $2^{(20°C/10°C)} = 4000$ h, and the two capacitors in parallel are thus about 2000 h. This is only about 3 months of continuous operation!

To increase the lifetime, we could switch to a 2000 h 105 °C rated capacitor. The lifetime is now 2000 h × $2^{(40°C/10°C)} = 32,000$ h, and the two in parallel are thus about 16,000 h. This is two years of continuous operation, and may be satisfactory for some applications. Of course, the increased specification of the capacitor comes at an increased cost as well. You can also get some additional lifetime by raising the voltage of the capacitor. For example, a 50 V capacitor in a circuit whose average voltage is 30 V will have 5/3 more lifetime. Of course, you still have to include the lifetime of the LEDs and other drive components to calculate the real lifetime of the system.

The other factor that sometimes has to be considered is the ripple current rating of the capacitor. The rated life of the part is also specified at a certain ripple current. For example, the 10 μF, 35 V electrolytic above has a rating of 1000 h at 85 °C at 36 mA. As temperature decreases, the ripple current rating goes up, but only slightly. At 65 °C, the ripple is rated at 45 mA. So generally, you don't get much of a bonus for running at lower ripple current than the rating. But you do have to ensure that the ripple current that the capacitor sees is not more than what it's rated for. The best way to check this is to actually put it in the circuit, and measure it with a current probe and an RMS meter rated for the frequency of operation.

A better choice for this and many LED DC drive circuits is to simply avoid the use of electrolytic capacitors entirely. For example, a 10 μF, 35 V ceramic capacitor is available in a 1206 package for a 2000 piece price of 6 ¢. In fact, a 10 μF, 35 V, 2000 h at 105 °C electrolytic runs 8 ¢ in a 5 × 6 mm package, and so the ceramic is a better choice anyway.

## ARRAYS

Most LED drive circuits use a single series string of LEDs. Since all of the LED forward voltages are in series, this sets the output voltage of the circuit and together with the input voltage determines the topology. But there is a downside to having all of the LEDs in series. If any one of them fails to open, then the entire string is open and there is no light at all. This is evident from the calculation of the lifetime of the string. Since all of the lifetimes are in parallel, the lifetime of the string is the lifetime of an individual LED divided by the number of LEDs.

An alternative is to put multiple strings in parallel. If one LED fails, taking out its string, the other strings continue to produce light. Although the total light from the circuit with this failure may be reduced, the reduction may not be enough to call the system failed. For example, if there are 10 strings and 1 fails, there is a 10% reduction

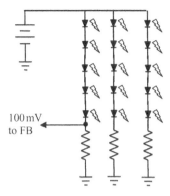

Figure 7.10   The effect of the current sense resistor is compensated by putting one in series with each string.

in light output. If total light is allowed to drop 30%, then this system still has a considerable amount of life left in it.

Simply putting multiple strings of LEDs in series doesn't work all that well. The problem is that the current sense resistor on the one string will generate a voltage mismatch with the other strings. A first pass at alleviating this is to put a resistor equal in value to the current sense resistor in series with each string, as shown in Figure 7.10. But even with 100 mV binning, a string of five LEDs can still have a variation of 500 mV from the next string. A way to deal with this is to have for each string a resistor in series that drops about 500 mV, as shown in Figure 7.11. The variation in voltage will then be partially taken up by the resistor, and the currents will be better balanced (confusingly, a resistor used this way is the original meaning of the word "ballast"). The resistor works better than the LED because a big change of current results in a big change of voltage across the resistor, unlike the LED where a big change of current is required to obtain just a small change in forward voltage. If the control IC requires less than 500 mV, it can be divided down.

The problem with these schemes is that if the one string being used to sense the feedback voltage fails open, there is no feedback at all, and something unfortunate may happen to the circuit. You could instead add all of the currents

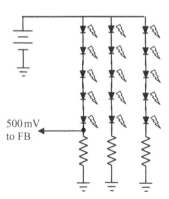

Figure 7.11   LED forward voltage variation can be compensated at the cost of additional power.

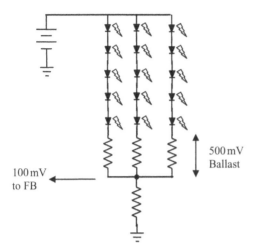

100 mV
to FB

500 mV
Ballast

Figure 7.12   Ballasting LED strings
with total current sensing.

into a single sense resistor, as in Figure 7.12. This way, if any of the strings fail open, the rest will remain powered. Of course, there is also additional power loss in this circuit, namely, the string current times 500 mV times the number of strings. Since this can be several watts, arrays are less common in practice than a single string of LEDs.

# CHAPTER 8

# PRACTICAL AC DRIVE CIRCUITRY FOR LEDs

**C**HAPTER 7 addressed DC drive circuitry for LEDs. This one is about AC drive. When your LED light needs to run off the electrical grid, AC drive circuitry is used. AC power conversion is considerably more complicated than DC. Not only is it technically harder, there are also government regulations concerning EMI emissions. This chapter will cover in depth what you need to know to be successful. The very first thing you have to know about is your *safety*.

## SAFETY

Did your parents tell you not to stick that paperclip into the wall outlet? Or about not running a toaster in the bathtub? Your parents were on the right track. AC is potentially deadly. Before you do anything with AC power, you need to read this section and take it to heart.

One of the authors once watched as an engineer working on a 277 VAC lighting ballast accidentally touched the input line. The engineer was knocked backward off his chair, and lay twitching on the floor for seconds. This guy was lucky. He didn't remain connected to the power line, and this probably saved his life. The next day he was back at work. However, there's no reason to count on your being equally lucky! (As an aside, you might wonder what *we* were doing. At first everyone was so surprised that we just stared. Then we ran around turning off the power. The authors strongly recommend that you get a "kill switch" installed in your lab before you work on AC power. This is a big red button near the doorway that you activate by hitting it with your fist. It should turn off all electrical outlets in the lab, but not the light. That way if there is an accident, you don't have to scramble around trying to figure out where the power switches are.)

The most serious danger from AC power is fibrillation, that is, your heart muscles twitching uncontrollably. Suppose you touch one side of the AC line with one hand, and the other side with the other hand. Now there is a path for current to flow through your body. And since your heart is in that path, it may receive enough of a shock to stop working.

*Practical Lighting Design with LEDs*, Second Edition. Ron Lenk and Carol Lenk.
© 2017 by The Institute of Electrical and Electronics Engineers, Inc. Published 2017 by John Wiley & Sons, Inc.

Another path is through your feet to the ground. If you touch the AC line with your hand, current may flow through your body and your shoes to earth. This can also potentially stop your heart.

It is worthwhile noting that Underwriters' Laboratories (UL) has done tests on AC safety. They have determined that even 5 mA is enough to create an unpleasant shock. At their safety class, they let you try out 5 mA, then 10 mA, and finally 20 mA—apparently very few people dare go to 20 mA after feeling the 10 mA shock! At 120 VAC, even a 24 kΩ resistor between you and line voltage is not enough to protect you.

Here is a series of recommendations to avoid getting into trouble:

1. Switch off the AC line before doing anything on a circuit. Suppose you have a probe looking at the source of a transistor, and you want to look at the drain. Turn off the AC power first, then move the probe, and then turn the circuit back on. Do this even with low-voltage points in the circuit such as the source. You never know if you're going to accidentally bump the high voltage. This seems like a nuisance in the lab, but your safety should be important to you.

2. Don't take someone else's word that power is off. Sometimes you will be working with someone else on an AC circuit. You've been debugging, and he tells you to take a look at something. Stop! You need to personally look and see that the input voltage monitor reads zero or, better yet, that the plug is unplugged.

3. Use an isolation transformer. An isolation transformer is a big, heavy transformer made of steel. It provides isolation between the input AC line and its output. The output is usually 120 VAC, but sometimes it is double this. The important part is that if for some reason the output is shorted, its power is limited by saturation of the core. You can't get more than 50 W, or whatever its rating is, out of it. A 50 VA device, suitable for much LED work, weighs 3 or 4 pounds. If it is much lighter than this, it's probably not the right device.

4. Use a fuse in your circuit. While you can still get a lot of power through a fuse, it's time limited. You won't stay connected to the AC line forever. Of course, only components downstream of the fuse are protected.

5. Keep one hand behind your back if you have to touch a live circuit. Since the worst danger is current through your heart, if one hand is behind your back, your body won't create a circuit from one arm, across your chest, and out through the other arm. It may be also convenient to grab hold of your belt.

6. No cheaters. We're not talking about copying someone else's design. A cheater is an AC adapter that converts a three-prong plug to two prongs. They may give the impression of offering isolation, but they don't. That third prong is there for a reason, to give you a ground connection. Discarding it may allow the power to float up, presenting a safety hazard. We've also seen an oscilloscope with a cheater have its probes literally burn when they were attached to power. Our suggestion: Ban cheaters from the lab.

7. Use insulation. This is the simplest and most obvious safety measure. After you finish wiring something with AC, wrap it up in electrical tape. (Ordinary masking or cellophane tape isn't good enough; it needs to withstand high voltage.) If

you're going to leave a probe on a point for a while, wrap the tip of it up in tape too.

8. Put up a sign. Nobody except you really knows if a circuit is live or not. Put up a big "Danger! Live 120 VAC!" sign so that people coming through know to not touch.

9. Turn it off. If you're going out of the area for a bit, to the restroom or even to get a component, turn off the circuit. Maybe the vice president of the company will come by showing a customer the work you're doing. Electrocuting the customer is bad for business.

10. Cordon off the area. If you need to leave an AC circuit on overnight, ensure that it is very safe. The cleaning crew at night needs to be safe too—try adding skull and crossbones to your sign, and a language or two other than English. Put up a plastic chain-link fence that someone has to actively push aside to get to the circuit.

Even with the best advice, there's no such thing as complete safety with AC power. Paying attention to what you and others are doing is your best safeguard. As a final note, you might wonder about that toaster. The reality is that in most modern bathrooms, there is a GFI (ground fault interrupter). This detects if there's any current going to ground rather than going to neutral. If there is, then it shuts off the AC power. So you're actually probably safe to make toast in the bathtub, if the house is new enough to have a GFI—but it's a good idea not to try it anyway.

## WHICH AC?

We say "the AC line," but this is shorthand for a very complex electrical environment. To design in this environment, you have to know what it's like. To start with, what voltage is the AC line? If you're in the United States, you probably will say 120 VAC. But in the rest of the world, it's usually 240 VAC. And 120 VAC isn't necessarily the right answer even in the United States. There's also 208 VAC, usually used in homes for dishwashers and the like, and 480 VAC used in industrial settings. While these two probably won't concern you as a lighting designer, there's also 277 VAC that is commonly used to power fluorescent ballast fixtures. This will be important if you're designing LED replacements for this type of lighting.

The 120 VAC line isn't really 120 VAC. That is its mean value. In fact, the power company usually (but not always) guarantees that the line voltage will be 120 VAC ± 10%, which is 108–132 VAC. This still isn't good enough for a design. If your customer is at the end of a long transmission line, the range will be more like 85–135 VAC.

This still isn't good enough. Sometimes there are brownouts. For one reason or another, the line voltage goes lower than 85 VAC. If it sits there for any length of time, motors tend to burn out, but your lighting shouldn't. After all, incandescent light bulbs are just resistors and they don't fail with low voltage, they just become dim. So at the least, your LED circuit should be able to survive any line voltage from 0 to 135 VAC indefinitely. It doesn't have to work properly at voltages outside the normal range, but at least it shouldn't fail. Testing your light by letting it sit at 10 V intervals for an hour each from 0to 135 VAC should be a required design test for your design.

Designing for the range of 0–135 VAC covers you for the U.S. market. But in the rest of the world, 240 VAC is widespread. And this has the same tolerance, so you need to design to 265 VAC. Incandescent light bulbs are country specific, so you don't necessarily have to design for both 120 and 240 VAC, normally called universal input. If you do, then you should have no problems, at least as long as the circuit produces the right light at both line voltages. If you don't design for universal input, you face a problem in Japan. There, half the country is on 240 VAC, the other half is 100 VAC (there are historical reasons for this).

In summary, deciding on the input voltage range of the driver is not just a technical decision. It also has market impact. The engineering and marketing departments have to decide on the following:

1. Will there be universal input, which costs extra money for each unit made? Or country-specific input, which increases the number of different designs for both engineering and production?

2. Will the unit produce the same light over some range of input voltages, and, if so, what is that range? At what voltage will it turn off? Or will it imitate incandescent light bulbs and have light dependent on input voltage? If the latter, will it be linearly dependent, or quadratic? (Remember that an incandescent light bulb is a resistor, so its power is represented as $V^2/R$, quadratic in the input voltage to first approximation.)

There are yet more things to consider about the AC line. What is the voltage when you're connected to 120 VAC? (Yes, this is a trick question.) Normally, you can assume that the voltage that the power company provides is sinusoidal. In this case, the voltage varies each half-cycle from 0 V to $120\sqrt{2} = 168$ V, the peak value of the $120\,V_{rms}$ line. But the factor of $\sqrt{2}$ is right only if the voltage is perfectly sinusoidal. In reality, there may be heavy loads on the line that change this.

How can the AC line be affected? Remember that what the power company generates is far away at a substation. It then goes through some long wires that may include other things, such as big motors. These long wires have impedance, and so their currents can change the waveform. Even inside a house, the wiring gives nonzero source impedance. For example, if you pull a lot of current (for example, turning on the microwave), you get some drop in voltage. If you momentarily feed a lot of current back (turning off the motor in your vacuum), you get some surge in voltage. The $120\,V_{rms}$ line can have both sags and spikes on it.

Then there is lightning. When lightning strikes near power lines, it can induce high voltages and currents on the AC line. This surge can produce voltage as high as 6000 V or current as high as 3000 A inside a house! The limits are set by arcing in the house wiring. Fortunately, these enormous voltages and currents are present only for very short times. We'll cover how to protect against lightning a little further on.

Before we close this section, let's briefly mention AC line frequency. In the United States, AC voltage runs at 60 Hz; elsewhere it is frequently 50 Hz. These frequencies are fairly precise, since they are related to the speed of the generator. Still, the precision in this measurement is for the long-term average frequency, not in an isolated moment. For most lighting applications, this won't matter, but it is traditional to design AC circuitry to work at 47–63 Hz.

## RECTIFICATION

A block diagram of an SMPS for running LEDs off-line is shown in Figure 8.1. The first step in using AC power (aside from EMI and surge protection, which we will cover below) is rectification. Rectification converts the sinusoidal AC line to DC with some ripple (see Fig. 8.2 for a bridge rectifier). When hot is positive with respect to neutral, current flows through diode D1, through the load (here a resistor), to ground, and then through diode D3 from ground back to the neutral. When it's negative, neutral is positive with respect to hot. Then the current flows through D2, *through the load in the same direction*, through D4, and then back to hot. Since current flows through the load in the same direction regardless of the polarity of the AC line, this is a DC supply. Of course, the actual voltage the load sees still runs from zero to peak each half-cycle.

Now there are a few options with the rectification. The one shown in Figure 8.2 is by far the most common. Note that the diodes are regular rectifiers, not fast or Schottky diodes. At 60 Hz, the recovery time of the diodes is unimportant (actually, regular diodes are better than fast because of EMI). And at 120 V, a 2 V forward drop (a pair of diodes conducting at the same time) is not a very significant power loss. The real choice is between discrete diodes and a bridge in a four-pin package. This choice is usually based on cost and space, the integrated bridge having smaller footprint but costing a bit more.

There are two more choices for rectification that might be possible for LED lighting. For high-power designs (in the multi-kilowatt range), the loss in the bridge, although a small percentage of the total power, may still be too large as a number of watts. Sometimes, MOSFETs are used to replace the diodes. When they're on, their

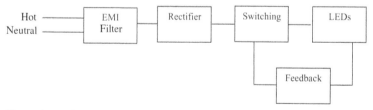

Figure 8.1    Block diagram of AC SMPS for LED lighting.

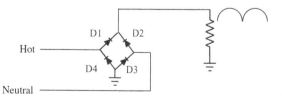

Figure 8.2    A bridge rectifier.

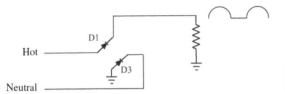

Figure 8.3    Half-wave rectification.

loss is $I^2R$, which may be lower than $I \cdot V$; when they're off, they don't conduct. To accomplish this, the drain acts as the cathode and the source as the anode. But driving the gate properly can be a little tricky. Since this high-power supply is expected to be unusual for LED drives, we won't get into details.

The other possible choice is half-wave rectification (the bridge in Figure 8.2 does full-wave rectification). See Figure 8.3 for how this works. When hot is positive with respect to the neutral, it works just the same as does full-wave rectification described above. Current goes through diode D1, through the load to ground, and then through diode D3 back to the neutral. But during the part of the cycle where the hot is negative with respect to neutral, there is no conduction. There is no output from the bridge. This system works in that you get DC. Current flows through the load in only one direction. But it is usually applied only at very low power, since it makes a mess of the AC line. It may also require more capacitance to support the load during the nonconduction time. And after all, for 2 ¢ more you can get the two extra diodes.

Now that we've rectified the AC line, the next question is whether we like the variation in voltage. Just rectifying the 120 VAC produces an output that goes between 0 and 168 V every 8.3 ms (1/60 Hz, and twice each cycle). This is undesirable because of visual flicker and potential effects on LED lifetime. The magnitude of the variation could be reduced by adding a capacitor, as shown in Figure 8.4. How much capacitance is needed? We can determine this by looking at energy stored in the capacitor.

As a side note, whether or not the light is flickering depends on who you ask. Some people can see 60 Hz flashes from fluorescent tubes. Others won't notice anything even at 50 Hz. What about the rectified 120 Hz? It turns out that in the real world rectified AC is slightly asymmetric, causing a 60 Hz modulation on the 120 Hz flicker. So the solution is to reduce the amplitude, making the flicker even less noticeable.

Energy stored in a capacitor is $(1/2)CV^2$. Now at the peak of the line cycle, the capacitor will be charged up to this peak, 168 V minus two diode drops, 166 V. So the energy stored will be 13.8 C in mJ when C is given in µF. Suppose that you want the minimum voltage to be 10% less, or 149 V. The energy stored in this condition is 11.1 C mJ. So the energy taken from the capacitor is 13.8 C − 11.1 C = 2.7 C in mJ when C is in µF.

Now we do a trick. If you look at the waveform in Figure 8.4, you see that after the peak is reached, the bridge input goes back down to zero, and then comes back up

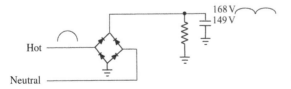

Figure 8.4    Reducing the
ripple from a bridge rectifier
with a capacitor.

to that 149 V point. But how long that takes involves the sine function. Rather than do the math, we will say that the capacitor has to provide power during the whole time from one peak to the next, 8.3 ms. This is a good approximation for a capacitor whose voltage doesn't vary too much, and not a bad one even if it does. It's not off by more than a factor of 2 in the worst case.

How much energy is needed? If your supply draws constant power, it's easy. Suppose you have a 4 W input power. Then in 8.3 ms, you need 4 W· 8.3 ms = 33 mJ. We needed 2.7 C in mJ, and so we need 33 mJ/2.7 mJ C = 12.2 µF. We round it up to 15 µF to account for tolerances.

That isn't all that much capacitance. There are problems with power factor doing this (see the section below), but the real problem is the voltage. You need a 200 V capacitor at 15 µF. Your only choice is going to be an aluminum electrolytic capacitor. This capacitor at this voltage is relatively large (such as 21 × 38 mm). And for a 240 VAC system, a suitably rated capacitor may not be even available. Below, we'll give some guidelines on how to deal with this. As we'll also see below, this type of capacitor may also determine the lifetime of your system.

Now there's one more choice for a drive. Certain manufacturers, notably Seoul Semiconductor, make "AC LEDs." These are LED devices in a single package that have a forward voltage of something like line voltage in both directions. They are intended to be run directly from the AC line, without any other components, not even the bridge in Figure 8.2. Now, one could imagine there are going to be some problems. What happens when the line voltage increases? The current of course goes up, and the manufacturer says that's acceptable. But you still have to dissipate all that extra heat somehow. And what happens when there's a line surge or lightning? It would seem you need some protection circuitry in front. And then there's EMI. We haven't talked about it yet (see below), but fast diodes such as LEDs can generate a lot of electrical noise. Perhaps an EMI filter will be necessary. The authors think there's more work to be done on this structure before it's ready for prime time. But if these issues can be worked out, then the AC drive circuitry discussed may well disappear. Ultimately, it's going to come down to cost.

## TOPOLOGY SELECTION

Now that we have rectified the AC, we need to do something with it. The general structure (the "switching" block in Figure 8.1) of the power supply is referred to as its topology. There are a number of possibilities. Let us start by understanding the pitfalls of the simpler, nonswitching topologies.

Analogous to what we mentioned above, the easiest thing that you could do would be to simply attach a suitable number of LEDs in series to get a forward voltage approximately equal to the AC line voltage (see Fig. 8.5). This is certainly very cheap. But it has the same practical problems as the integrated version. The line voltage goes from 0 V up to 168 V and back down to 0 V 120 times per second. While this isn't visible to most of us, some people can see it (especially if it's 100 times per second rather than 120), and it's very hard on the LEDs. This variation can be fixed with a capacitor, as we discussed above. But worse yet, when the line voltage goes up 10% to

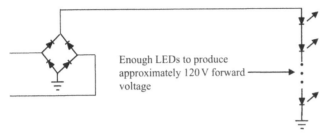

Figure 8.5   Running LEDs directly off-line.

132 V, there is a *lot* more current into the LEDs. Remember that they have current that is exponential in the voltage, much faster than the quadratic rise of incandescent light bulbs. And when the line voltage goes down 10% to 108 V, they're going to be quite dim. In short, while series connection of the LEDs may possibly be acceptable when you have a fixed DC voltage, the AC line varies too much for it to be practical in real off-line applications.

## Definition of Ballast

The word "ballast" is used in two different ways. Traditionally, a ballast is a resistor used in the emitter of a transistor to compensate variations in voltage that affect gain. From there, it was straightforward to use the word ballast to signify a resistor in series with a string of LEDs. However, ballast is also the word used to mean the circuit that drives a fluorescent tube. This is because the original circuit used a capacitor to limit the amount of current into the tube. When electronic circuits for fluorescents became available, they were also called ballasts. In this book, we often call circuits for running LED "ballasts" also. But you should be aware that some people take exception to this, and call such circuits "drivers." For our purposes, there are no differences between the two terms.

A slightly more complex choice is to linearly ballast one or more strings of LEDs with resistors. Imagine putting a resistor in series with the LED string in Figure 8.5. Now this ballast resistor is supposed to take up the variation in the line voltage. If the voltage becomes too high, the current through the LEDs goes up. This causes the voltage across the resistor to also increase. Thus, the resistor absorbs part of the increase in input voltage. But this comes at a cost. Suppose normal current is 40 mA, which is $120\,V \cdot 40\,mA = 5\,W$ into the LEDs. If you want to be able to take up $10\% = 12\,V$ on the resistor, the resistor value should be $12\,V/40\,mA = 300\,\Omega$. In normal operation, this resistor dissipates $(12\,V)^2/300\,\Omega = 480\,mW$. But now suppose the voltage rises to 132 V. The LED voltage doesn't change much with current, so almost the whole increase in voltage goes across the resistor. Power in the resistor is now $(12\,V + 12\,V)^2/300\,\Omega = 1.92\,W$. Since the input voltage is capable of sitting at this voltage indefinitely, you need a 5 W resistor. This is a $1.5\,cm \times 1.6\,cm$ package. And you need to be able to dissipate an extra 2 W thermally. At an overall electrical efficiency of $(7 - 2\,W)/7\,W = 71\%$, a resistor is not a practical solution.

A better way to deal with line voltage variation is to use a linear regulator. You could imagine a voltage regulator that ran a string of LEDs at the same voltage regardless of input line voltage. You could even notice that as the LEDs warm up, their forward voltage and efficacy both decrease. So with a cleverly designed voltage regulator, this system would produce constant light output: As the forward voltage drops, the current increases, compensating for the efficacy decrease. Unfortunately, high-voltage input regulators don't currently exist.

The other way to run LEDs with a regulator is to use one with a current output, producing the same current into the LEDs regardless of input voltage. There are such devices on the market. You could even make an additionally complex one that compensates for forward voltage drop by increasing the current.

The problem with both of these linear regulator concepts is that they still dissipate too much power. Any increase in line voltage is dropped across the linear regulator. So at 40 mA and 108 V across the LEDs, a 120 V input loses $(120 - 108\,\text{V})\cdot$ 40 mA = 500 mW in the regulator, and when line goes up to 132 V, 1 W. So the regulator has to be pretty strong to handle so much heat dissipation.

Since these linear schemes are out for most applications, we will use a switch-mode power supply. We have available the various topologies mentioned in Chapter 7 on DC drives, although there will be additional choices once we've added a transformer for isolation. Here we'll talk about the nonisolated types of converters for off-line use and defer the question of isolation to the next section.

As in the previous chapter, we will be using a switch-mode power supply that regulates current. The easiest of these to work with off-line is the buck type, as epitomized by the 9910 IC (parts by ST and others work much the same way). As you recall from the previous chapter, the buck is characterized by producing voltages that are strictly less than or equal to the input voltage. Since the input here is 168 V and LEDs are only a few volts each, the buck is almost always a good choice.

Take a look at Figure 8.6 to understand conceptually how the off-line buck works. When MOSFET Q1 turns on, inductor L1 is effectively grounded. So the input voltage drops across the series LEDs, and then is dropped across the inductor, which

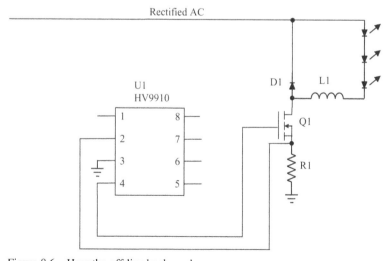

Figure 8.6   How the off-line buck works.

starts to increase its current. Of course, since they're in series, the inductor current is the same as the LED current. When the current reaches a specified level (set by the IC and sensed by R1), the MOSFET turns off. Now the current in the inductor can't go through the MOSFET, and it has to go somewhere, so it goes through the diode D1, back to the input. The voltage on the end of the inductor that used to be grounded is now close to the input voltage, while the other end of the inductor remains lower by an amount equal to the LED voltage. So in this part of the cycle the inductor current and the LED current both decrease.

By controlling the point at which the MOSFET turns off, the current can be maintained at approximately a desired value. And if the inductor is relatively large, the minimum current will be close to the maximum current, so that an approximately constant current runs through the LEDs. This circuit also has the convenient feature that both the diode and the MOSFET never see a voltage higher than input voltage, just like other buck converters. This makes component selection relatively easy.

We might note that you could use the 9910 to generate an off-line buck-boost, just as in Chapter 7. However, there would be no point in this, unless you had an exceptionally long string of LEDs. Then a buck could only be driving them during the fraction of the time when the line voltage was higher than the string voltage. For most applications, the LED string voltage will be a fraction of the peak line voltage, and the buck version will work fine.

## NONISOLATED CIRCUITRY

A real circuit using the 9910 is shown in Figure 8.7 (minus the input filter). We will be using this circuit in a BR40 design in Chapter 13. For the time being we'll do a quick overview. The components in the schematic that haven't been mentioned yet are

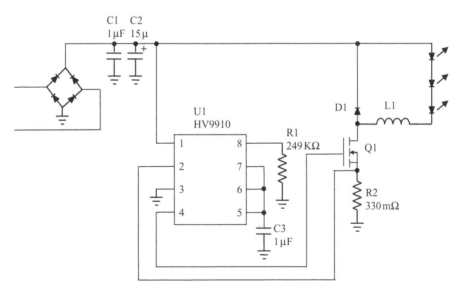

Figure 8.7 A nonisolated off-line LED driver.

related to control of the converter. Resistor R2 sets the current. When the voltage at R2 reaches 250 mV, the MOSFET Q1 is turned off. Turn off is thus at 250 mV/ 330 mΩ = 758 mA. Capacitors C1 and C2 are bypass for the switching (they both experience line voltage and must be rated for 200 V). C3 is bypass for the internal linear regulator, which is 7.5 V.

Resistor R1 sets the switching frequency of the converter. A higher frequency means a smaller inductor. So generally, higher is better. But a higher switching frequency may mean it's harder to filter the EMI (see below). A practical compromise between these conflicting desires is to pick a switching frequency just below 150 kHz. Since EMI requirements start at 150 kHz, this puts the main switching power where it doesn't need to be filtered. R1 at 249 kΩ produces a switching frequency at 90 kHz.

Now look at the power portion of the circuit. Transistor Q1 switches on and off at 90 kHz under the control of U1. Again, U1 is controlling the peak current measured through R2. When Q1 is on, the current flows through the three LEDs, through the inductor, and then through the transistor and R2 to ground. Since there is a positive voltage across L1, its current increases during this time. When it has increased enough to reach the threshold set by R2, the transistor is turned off. Now the current through the LEDs and L1 is shunted through D1 back to the input. Since the voltage across L1 is now negative (one side is down three LEDs and the other side is up one diode), the current decreases. Then after 11 μs (one over 90 kHz), the whole cycle starts again.

This is an extremely simple power supply, and that is its advantage. There's no compensation circuitry design necessary, there's not much in the way of magnetic design, and there are only eight pins on the IC. If a nonisolated supply is acceptable for your purposes, power supply design doesn't get any easier than this!

## ISOLATION

You'll recall from this chapter's introductory section on safety that isolating the system from AC line with an isolation transformer is one of the important safety techniques to be used in the lab. Isolation is also very useful for products for the same reason, although it's not done the same way. A customer must not accidentally touch something that is connected to live AC. And Underwriters' Laboratories standards require isolated power supplies in many cases. Some further information on these requirements is in the section below on UL. (Note that we are using "UL" to refer to generic safety standards. Underwriters' Laboratories writes these standards with input from a variety of parties, but there are a number of companies that will test devices to these standards, UL being only one of them.)

The sort of isolation transformer you use in the lab is large. A 50 VA unit can weigh 5 pounds. Clearly, this will not work for a lighting product. The reason that the isolation transformer is so big is that it is transforming 60 Hz (or 50 Hz). The low frequency means that the voltage is positive for a long time before it becomes negative. The inductor therefore has to be large to support all of those volt-seconds.

In an SMPS, we can employ a trick. Since it is switching at high frequency, at least 20 kHz and often 10–100 times higher than that, the volt-seconds requirement for the transformer is proportionately reduced. You can transfer substantial power through an isolation transformer that is not substantially bigger than some capacitors. You simply put the transformer in place of the inductor that is there anyway, and separate off the load from the primary. With a properly designed transformer, everything on the secondary side is then UL safe.

That said, however, you should be aware that many designs use nonisolated converters. Nonisolated converters are both cheaper and easier to design than isolated ones. In a real product, after all, the converter is going to be inside an electrically insulated metal can or a plastic or glass enclosure of some kind. The customer shouldn't be able to access the inside of that. So then the only AC live parts that are potentially exposed are the LEDs and their PCB traces. If these are securely doubly insulated, then a nonisolated supply may be acceptable. Again, we recommend you follow UL standards and apply for UL certification.

Now how do we insert a transformer into Figure 8.7 in place of the inductor? There are two choices, shown in Figures 8.8 and 8.9. Looking first at the schematic in Figure 8.8, we see that it is a forward converter.[1] You can tell that it is related to a buck because when the transistor turns on, power is transferred to the load, the same as a buck. It can therefore be controlled by the same IC that controls a buck, and so might be a good choice. But the downside is that whereas the buck had just one diode, this converter requires two diodes and an inductor on the secondary side. At the high currents required by LEDs, this inductor will be large, and so the forward converter is probably not the right choice.

Turning to the schematic of Figure 8.9, we have a flyback converter.[2] You can tell that it is related to a boost because when the transistor turns off, power is transferred to the load, the same as a boost. The difference between the flyback and the forward transformer is just the polarity of the transformer, signified by the dot. This converter has the same number of components as the boost, there is no extra inductor and only one diode on the secondary, and so is a good choice. Normally, special measures would have to be taken to have an IC control the flyback. But in this particular case, we are controlling the peak current cycle-by-cycle. It will turn out that

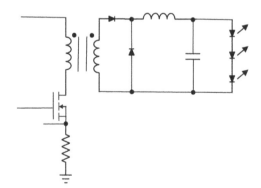

Figure 8.8    Adding a transformer makes the converter into a forward.

---

[1] A forward is a buck-derived converter. For more information on this converter, refer to Lenk (1998).

[2] This is a boost-derived converter.

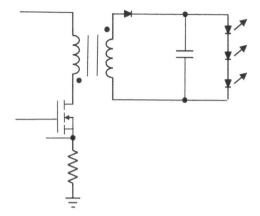

Figure 8.9    Adding a transformer
makes the converter into a flyback.

in this case, the same IC that we used for the original buck can also be used for this flyback, without raising concerns about loop stability.

Having chosen a flyback topology for our off-line converter, what we haven't yet said is that the next step is to decide whether this should be a continuous or discontinuous mode flyback (see Lenk (1998) for more information). There are pros and cons to both.

1. *Transformer size:* In a discontinuous mode flyback, the transformer acts as an inductor during the transistor on-time. All of the energy for the cycle is stored in the primary inductance. This results in a relatively large transformer. In a continuous mode flyback, however, the transformer is just a transformer. Therefore, it is rather smaller than in discontinuous mode.

2. *Peak LED Current:* In the discontinuous mode flyback, no energy goes into the capacitor while the transistor is on. Thus, to get the same average current, the current, when it does flow, must be higher. This means that the capacitor must be adequately sized for the ripple current, which may add to the size problem. The continuous mode flyback has current going through the capacitor all the time and so peak and average currents are almost identical.

3. *Control Signal Isolation:* The LEDs are on the other side of the isolation barrier from the controller. Somehow the IC has to detect how much current the LEDs are getting, so the gate drive is turned off at the right time. Now in a discontinuous flyback, we get a lucky break. When the transistor is on, all of the energy is stored in the primary side of the transformer, which acts as an inductor. When the transistor turns off, all of the stored energy goes to the secondary side so that, on average, control of the turn-off point of the transistor determines the LED current.

   The continuous mode flyback is more complex. There is always current coming out through the secondary of the transformer, and information on the amount of current is available only during the time the transistor is on. In short, the IC isn't taking into account the duty cycle. Information about the LED current has to be transferred across the isolation barrier. This requires additionally circuitry, usually fairly complex, expensive, and prone to failure circuitry. For example, the common way to do secondary side current sensing is to use a 431 voltage reference IC as a comparison for the current sense voltage, and then use it to control an opto-isolator. The opto-isolator then controls the feedback of the IC. But the opto-isolator has a number of potential problems, including poorly controlled gain and aging.

Given this last problem with continuous-mode flyback converters, we usually choose a discontinuous mode flyback for our isolated converters. These have some limits on how much power they can easily transfer, but for a 120 VAC line, this is well over 50 W. If you're going to build a light that uses more power than that for the LEDs, you probably can afford the size and cost of a larger ballast anyway.

## COMPONENT SELECTION

One issue that hasn't been addressed in the previous two sections is how to select the values of the components. Some of it is obvious. If the LEDs conduct 400 mA, you don't want to use a 500 mA diode, even though it would probably work. You want to use a higher current diode, to take advantage of a lower forward voltage. Similarly, for the MOSFET, you want to use one that has relatively low $R_{DS,on}$ so that your losses aren't too great. But what about the breakdown voltage, both for the MOSFET and diode, and for other components such as the input capacitor and the IC?

As a general rule, you want to obtain the highest voltage part you can afford. The peak voltage on a 120 VAC line running 10% high is 187 V, so a 400 V device gives plenty of margin. What do you need margin for? The input filter will include an MOV (metal-oxide varistor) for lightning protection. But the MOV isn't like a Zener diode that clamps at exactly its rated voltage. Its voltage increases substantially when lightning current is added to it. For example, the ROV07-241K sits at 240 V at 1 mA (where it is already dissipating 1/4 W), but at 10 A it's up to 395 V! Parts that are being protected by the MOV need quite a bit of voltage withstand capability above the rated MOV voltage. A 400 V part is adequate on a 120 VAC line. But 240 VAC is within normal bounds at a peak voltage of 373 V. Even 600 V is a little too close here; 1000 V would be desirable.

Looking first at the diodes (both the rectifier and the switching diode), there's very little difference in price between 400 and 600 V. So you should choose at least 600 V. But the reality is that in production volumes, diodes are one of the least expensive components. So generally, 1000 V parts are the right choice. This will then cover you adequately both in the United States and around the world.

The MOSFET is somewhat more challenging. Devices of 600 V are common, and should always be used. This is high enough for use on 120 VAC. But that doesn't leave all that much margin for lightning let-through from the MOV on a 240 VAC line. So you have to make a money decision with marketing involvement. Do you pay the extra money for a 1000 V MOSFET and have a single design? Or do you use these only for the non-U.S. market, and have two different designs, with the costs associated with two different product codes?

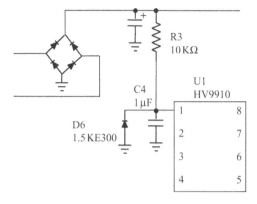

Figure 8.10  Protecting the HV9910 from high voltages.

The IC really ought to be rated at least 600 V. Although some are, in fact the 9910 is rated at only 450 V. This is probably adequate for the 120 VAC market, but is clearly unacceptable for the 240 VAC market. A suggested practical fix is shown in Figure 8.10. The resistor drops voltage equal to the current draw of the IC multiplied by the resistance. When the voltage surges above the TransZorb voltage, the current it has to absorb is limited by the resistor. A bypass capacitor is included. Shown are some typical values. This scheme doesn't cost much (the resistor and TransZorb are cheap) but does consume some power in steady state. And this is true for all the 240 VAC components in this section: 240 VAC circuits are going to be slightly less efficient than 120 VAC circuits. The reduced losses in the bridge are usually not high enough to affect this conclusion.

The most trying question is the voltage rating of the input capacitor. Aluminum electrolytic capacitors are almost nonexistent above 450 VDC. This is adequate for 120 VAC, but clearly unacceptable for 240 VAC. What can be done? It's well known that putting two capacitors in series halves the capacitance and doubles the voltage (opposite of a resistor). But capacitors have a tolerance that affects this division, typically ±20% for this type. If one of the two capacitors in series is 20% and the other 20% low, the one that is 20% low will take 20% more of the total voltage than the other one. This could exceed the rating of the component. The solution is shown in Figure 8.11. Adding large-value resistors in parallel with the capacitors helps to balance out the AC current. This ensures that voltages divide approximately evenly between the two caps even when their values are somewhat imbalanced. Having a resistor in parallel with the input capacitor may be a good idea anyway, since it ensures that the circuit will eventually dissipate its charge. Otherwise, the bridge

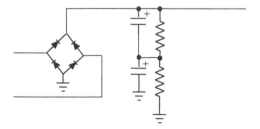

Figure 8.11  Resistors balance voltages for series capacitors.

blocks discharge of the capacitor in one direction, and the LEDs have too low a leakage current to discharge it in the other direction.

Still, a better plan is to not use an electrolytic capacitor at all, if that is possible. This is good for both power factor and lifetime (see below for both). The only capacitor needed is 10–100 nF of high-frequency capacitance for the high-frequency switching. Such capacitors are available at 630 V rating in an 1812 package, and at 1000 V in an 1825 package.

# EMI

One of the things that make design of AC ballasts much harder than DC is electromagnetic interference (EMI). Governments require that devices that attach to the AC line not produce more than a specific amount of electrical noise. And switch-mode power supplies, since they switch at high frequency, generate a lot of noise.

There are two types of noise, and one of the types has two varieties. Your design has to meet all of the noise requirements for all of these. The first type of noise to consider is conducted noise. This is noise that comes out of your design and is conducted along the input power wires. It is typically caused by devices that turn on and off quickly and carry power, in particular the MOSFET and the power diode. Devices that switch quickly but don't carry much current typically aren't as important for conducted EMI, although it's wise to create the best possible design just in case.

There are two types of conducted noise: normal mode and common mode. They are typically treated differently. Normal mode noise is due to different current in the hot and neutral power wires. Common mode noise is due to the current in the hot and neutral wires with respect to the ground wire. Note that for lighting purposes, many devices will have only two wires. In this case, only normal mode noise is possible, and that is the only type that can need filtering.[3] Normal mode filtering starts with a high-frequency bypass capacitor at the input to the switching stage (see Fig. 8.12), and will probably include a series inductor as well. Common mode filtering will require, in addition, a balun and capacitors from both hot and neutral to ground (Fig. 8.13). The balun can have normal-mode inductance as well as common-mode, so that a separate normal-mode inductor isn't required. There are UL limits on how great a value the two caps to ground can be.

The second type of noise is radiated. This is typically caused by current loops, that is, current that runs in one direction and then returns, with a physical separation between the two paths. An example of this is shown in Figure 8.14. The MOSFET pulls current from the capacitor through a trace, and the source of the MOSFET

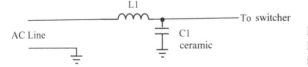

L1

AC Line

C1
ceramic

To switcher

Figure 8.12   Normal mode EMI filtering for a two-wire input.

---

[3] We see numerous designs that have only two-wire input and yet utilize a more expensive common-mode filter. This is seriously misguided.

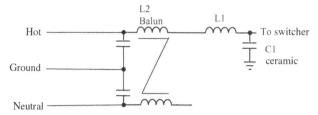

Figure 8.13 Common mode EMI filtering added for a three-wire input.

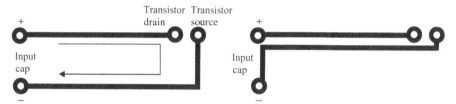

Figure 8.14 Current loops may cause EMI problems: reducing loop area helps.

returns the current to ground through another trace that is separated from the first trace. Current goes in a loop as shown, and this acts as a little antenna, radiating electromagnetic noise. The solution is to minimize the area of such loops, and try to return currents either on top or bottom of the forward path or, in the case of a single-layer board, next to the forward path.

One more concept to note is that a nonisolated supply should have an isolated heat sink so that customer contact with the AC can be avoided. This potentially worsens the thermal interface, but its EMI consequences are not as dire as some have made them out to be. In a power supply, the main radiators are the MOSFET and the power diode, which couple capacitively through the heat sink isolation to make the heat sink an EMI radiator. In an LED light, however, the components dissipating the most power are the LEDs. Since these by design don't have much ripple, they are not very important to EMI. And since the ballast is a comparatively small part of the power, it may not need to be heatsunk to the heat sink. The conclusion is again the same: an isolated supply may be easier to deal with.

It is beyond the scope of this book to describe how to get your AC converter to pass EMI. Detailed information on how this can be done can be found in Lenk (1998). Here are some pointers that will help you get there with a minimum of confusion:

1. Put all of the components close together on the PCB.
2. Avoid having large currents go in loops.
3. Don't pick components that switch faster than you need them to. For example, choose a normal rectifier for the bridge, not an ultrafast.
4. Have currents on traces that are as close as possible to their return traces, either above or below, or on a single-layer board next to the return trace.
5. Bypass the input to the converter with a small (<1 μF) capacitor. Small-value capacitors work at higher frequencies than large capacitors—the electrolytic in parallel with this small capacitor doesn't help the EMI at all.

We'll make a brief comment on EMI standards in the United States versus those in the European Union. These standards are fundamentally very similar. The only significant difference at this time is that the European Union also has a limitation on power line harmonics, EN61000-3-2. Chances are that if your design passes EMI, it probably will pass this standard as well.

Finally, let's comment on selecting the switching frequency of your SMPS. The reason for including these comments is that SMPS switching is often the most important source of EMI in a system. Now the fundamental consideration is that EMI-conducted limits start at 150 kHz. Below that frequency, you can have as much noise power as you like (subject to limits on harmonic content). Above 150 kHz, the amount of noise allowed decreases at 30 dB/decade, which is faster than a one-pole filter but slower than a two-pole filter. So the most sensible choices are either to switch below 150 kHz or else be considerably above that frequency. Switching at 200 kHz, for example, gives you the worst of both worlds.

You will also have to pass radiated emissions. The limits for this start at 30 MHz so that the emissions don't directly impact switching frequency selection. However, switching at high frequencies requires transitions between on and off in the power elements to be fast. If your switching frequency is 1 MHz, the on-time of the transistor is less than 1 μs, and so it needs to turn on and off in less than 30 ns to avoid big losses. These transitions are within the regulated radiated bands. If instead your switching frequency is 100 kHz, the transition times can be 10 times slower.

From an EMI perspective, a switching frequency below 150 kHz is going to be best. With lower frequencies, the downside is that component sizes get bigger. This is true particularly of inductors and capacitors. You can do a lot of math to find out that there's no one best answer; there are too many factors to consider and optimize. Our practical recommendation is this: pick a switching frequency just below 150 kHz (including component tolerances) or above 1 MHz. If size isn't an issue, the lower of these two is probably the better choice, and will be more efficient anyway.

## POWER FACTOR CORRECTION

Many devices above 50 W are now required to be power factor corrected (PFC). Additionally, new regulations (Energy Star in the United States, in particular) seem to require PFC for lighting devices regardless of power level. So your design may well need to be power factor corrected. What is PFC?

Consider the current input to a switch-mode power supply when there is a large capacitor right after the bridge (see Fig. 8.15). When the line voltage reaches its peak

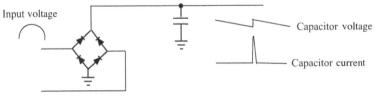

Figure 8.15    A big capacitor maintains constant voltage during the line cycle, generating large peak currents and bad power factor.

value, the capacitor is also charged to its peak. Now the line voltage decreases. But since the capacitor is large relative to the amount of energy the power supply is pulling from it, it discharges more slowly than the line voltage drops. Consequently, during this time, the line does not provide any current to the capacitor. Once the line has gone to zero and come back up, it reaches the same voltage as the capacitor, and begins to charge it again. During this short time, all of the current needed for the whole cycle is delivered to the capacitor. As a result, the current from the line has a big spike close to the peak of the line voltage. This spike is one of the primary causes of bad power factor.

Power factor is in a practical sense how close the input current matches the sinusoidal input voltage. What you want is to have input current also be sinusoidal, and in phase with the voltage. The load should "look like" a resistor. The current to the big capacitor clearly doesn't meet these criteria, and as a result has poor power factor.

What's wrong with a poor power factor? The reality is that it wouldn't hurt your design very much to have a bad PF (usually your input capacitor is rated to take the RMS current). It's rather a problem for the electric power company. When you have a load that doesn't look like a resistor, it must look like a capacitor (or inductor). These pull current without dissipating it. So the power company supplies you current, but you don't pay for it, since you return it on the next part of the line cycle, and this is costly to the company. You have power losses in the power lines because of this current, and the company needs additional generating capacity for something that isn't billable. To minimize these problems, governments mandate minimum PF for many devices.

There are a couple of methods to correct the power factor of your design. The traditional method is to add a second converter to the input of your power supply. In this type of design, the first stage is a boost converter that takes the input voltage and boosts it to a higher DC voltage, typically 400 V. It typically uses a special control IC that draws current from the line in phase with the input voltage, so that it has nearly perfect PF. Then the second converter takes this 400 VDC and converts it to the output, in this case an LED current. This two-stage converter works well, but does add substantially to the cost and size of the design.

A smaller and less costly alternative is shown in Figure 8.16. It basically replaces the single-input electrolytic capacitor with two capacitors and adds three additional diodes. It can probably work up to about 60 W. It is therefore a reasonable alternative to the two-stage design for many LED applications.

One of the authors devised an even cheaper scheme to do PFC with the 9910. This is shown in Figure 8.17. The brightness control pin of the IC is connected through a resistor divider to the AC line input. As the AC line goes up and down, so

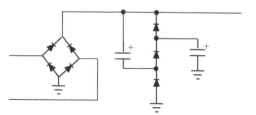

Figure 8.16    A smaller and cheaper PFC.

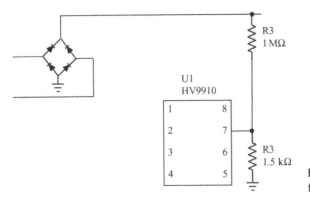

Figure 8.17    Simple power factor correction circuit.

does the input current. The top resistor of the divider is selected to give low power dissipation at 120 VAC; the bottom resistor is chosen to give full scale of the control pin (250 mV) at line peak (168 V). This circuit gives a PF of >0.9, and has essentially no cost and takes up no space. No electrolytic input capacitor is required, resulting in a cost saving and improving the lifetime of the power supply. The downside is that a sinusoidal current is now going through the LEDs, which as discussed below causes some loss of efficacy and potentially decreases their lifetime.

## LIGHTNING

Anything attached to the AC line will occasionally face lightning on the line. Before considering what to do about this (if anything), let's first look at the phenomenon. Lightning is a natural phenomenon, and so can vary wildly from event to event. IEEE587 (now ANSI/IEEE C62.41) defines the types of lightning events that are generally considered to be worst case. If you can survive them, then you're probably in good shape for the real world. A good overview with waveforms is given by Martzloff (1991), a pioneer in the field.

The first thing to know is that the type of lightning surge that occurs depends on location. Outside a house or a building, the surges are much worse than inside. Inside, the maximum voltage you can see is limited by the wiring. If the voltage goes too high, the wires will arc in the circuit box. Since most LED designs are done for indoor lighting, we'll concentrate the discussion on Class B.

Inside a building, the arcing limits the maximum voltage to 6000 V, and the wiring impedance limits the maximum current to 3000 A. These both don't occur at the same time. The source looks approximately like 6 kV with 2 Ω impedance. This still sounds like it's impossible to protect against them. What saves you is that the time during which this occurs is very short.

There are basically three types of lightning surges you will need to protect against. The 6 kV, 3 kA type has a rise time of 1.5 μs, and a decay time of 50 μs (and is thus called a 1.5/50 waveform). Then there is a 10/1000 waveform, but it has a peak voltage of only 1300 V and an impedance of 250 mΩ. Finally, there is a ring waveform, which rings up to 6 kV, 500 A with a frequency of 100 kHz, decaying

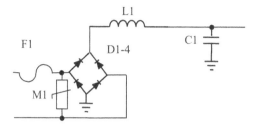

Figure 8.18    Adding an MOV to the design protects it moderately well from lightning.

away within a few cycles. There are other types of waveforms as well, but these three will practically cover your design.

What can be done to protect your 600 V design against 6000 V lightning? The best approach is to add an MOV to the input, as shown in Figure 8.18. An MOV resembles a large, power bidirectional zener. You can pump huge amounts of current into it for a short period of time, and it clamps the voltage. Because of this, it is a good device for protection against lightning.

However, there are several aspects of MOVs that make them less than perfect. For one thing, they have soft "knees." That is, their voltage is not absolutely fixed, but climbs up quite a bit as you increase the current. In practical terms, this means that you have to have a fair amount of margin between the turn-on voltage of the MOV and the voltage at which your converter will break. Furthermore, MOVs perform differently as they age. After a number of lightning events, their clamp voltage starts to rise. After enough events, they can explode, leaving your circuitry unprotected against the next lightning strike.

So the best you can do is to put in the biggest MOV you can. They are relatively inexpensive, so the determining factor is how much space your design has. For a 120 VAC design, a practical MOV is −241, rated for 150 VAC, and for 240 VAC −431, rated for 275 VAC.

Now if we are honest, we will recognize that many electronic designs don't bother with an MOV. Why don't they just blow up the first time there's a storm outside? Part of the answer is that they do—bottom-end electronic devices just fail after a while, and lightning *is* one of the reasons. Another consideration is that if there is a device on the circuit that is protected, it may protect other devices on that line as well. So some devices leach off others' better design practices. We recommend that you always include lightning protection, as LED lighting is advertised to last for a long time. It's bad if the LEDs are still working great, but the lamp is dead for the lack of a 10-cent part. And it's poor design practice to hope that someone else is protecting you. Ultimately, the bottom line is cost, which is a marketing decision.

## DIMMERS

There are many homes with dimmers for their light bulbs. They are also fairly common in businesses—restaurants dim lights for ambience, conference rooms have light dimmers for presentations, and so on. So at least for LED designs going into general service lighting, the ability to work on a dimmer circuit is highly desirable.

Figure 8.19   Output waveform of a triac dimmer.

The trouble is that dimmers are a problem for most AC power supplies. Almost all of them work by turning off the AC power during a portion of each line cycle. This is almost always done with a triac circuit, because it's amazingly cheap. And since it's such an inexpensive way of doing it, you should expect triac dimmers to continue to dominate the market in the foreseeable future. There are also electronic dimmers, but we won't consider them.

The basic waveform generated by a triac dimmer is shown in Figure 8.19. The voltage is zero until some phase angle is reached, and then it abruptly resumes the normal sinusoidal voltage for the rest of the cycle. The missing part can be either at the beginning of each cycle or at the end, called respectively leading or trailing edge. You see both kinds in commercially available dimmers.

There are really two problems with this waveform. The first is that when the line voltage abruptly turns on, there is a large surge of current into the input capacitor, which tries to charge up very quickly. This can cause an electrolytic capacitor to blow up within minutes. The solution here is to be PFC. Without the big capacitor, the rest of the circuitry doesn't mind the chopped AC waveform. It regulates the LED current during the portion of the cycle when the voltage is present, and the average LED brightness is reduced proportionally. This works as long as your converter's bandwidth is fast enough to respond to the missing part of the line voltage. (PFC does not require low bandwidth. For example, the gain circuit suggested in the section above will work just fine.) The only trick you may need to know is that if the IC takes a long time to turn on (some take milliseconds), then you may need to provide power to it during the line's off time. All this takes is a diode and a capacitor, as shown in Figure 8.20. The capacitor is chosen to provide current to the IC during one cycle of 8.3 ms, and the diode blocks the capacitor from discharging back to the line while the input is zero.

The second problem is more serious: LEDs are so efficient that the dimmer may not work! The reason is that dimmers are meant for incandescent light bulbs, the

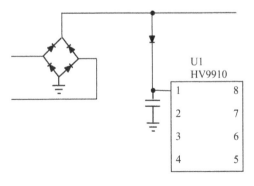

Figure 8.20   Keeping an IC's power alive during the off-time of a dimmer.

smallest of which is typically a 40 W device. Many LED light bulbs, however, are sub-10W. Below about 30 W, the triac doesn't work right, you end up with abnormal AC waveforms, and the whole system doesn't dim properly. This problem is not solved by PFC, and indeed is made worse by it, because the current drawn by the supply is lower at the lower line voltages.

One solution that some IC manufacturers have been implementing is to increase the load current. This could be accomplished by adding a big resistor at the input. But then the LED efficiency is reduced to that of an incandescent. Instead, the IC people attach the resistor close to the zero-crossing of the AC line. This is the portion of the cycle where the triac needs to have enough current, so the resistor ensures it fires properly. Additionally, since the line voltage is low during this time, the power dissipated in the resistor is relatively low. Undoubtedly, there is room for further design improvements in this developing field.

## RIPPLE CURRENT EFFECTS ON LEDs

The authors once built an array of LEDs and ran it with a lab supply set to current limit. The light was measured with an integrating sphere, and the current was set to produce sufficient light output. The current was then measured with a true-RMS meter. Next, an AC power supply was built and measured to have the same current. But when a light measurement was taken, the LEDs were dimmer. It wasn't a meter problem, because the meter had plenty of bandwidth. The problem, it was determined, was that LEDs lose efficacy when there's ripple current. So we had to use more power to obtain the same light output. We're going to examine the magnitude of this effect in this section. Some graphs showing approximate effects are shown in Figure 8.21.

There are actually two aspects of LED performance that are affected by ripple current that acts to increase power dissipation. To get the same light output from a

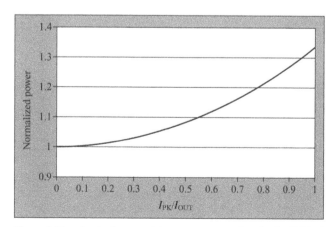

Figure 8.21   As ripple current increases, power loss in the LED also increases. (*Source:* Betten and Kollman (2007). Used by permission of Electronics Technology, a Penton Media publication.)

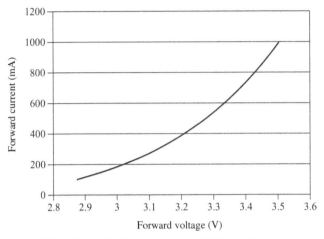

Figure 8.22 Forward voltage increases with increasing current.

rippling current as from a constant current requires that the peak current be higher than the average. Now recall that the forward voltage of an LED is only approximately constant with current. In reality, the forward voltage increases somewhat as the current increases. Looking at Figure 8.22,[4] we can see that increasing the current from 350 to 500 mA increases the forward voltage from about 3.18 to 3.28 V, about a 3% increase. This increases the power dissipated while the current is this high.

A second loss mechanism is that the efficacy of the LEDs decreases with increasing current. Looking at Figure 8.23, the relative flux at 350 mA is 1.00, while at

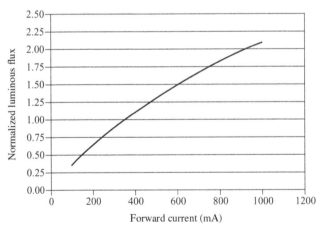

Figure 8.23 Increasing the current does not proportionally increase the light. (*Source:* Technical Datasheet DS56, Power Light Source Luxeon Rebel, Philips Lumileds Lighting Co., 2007.)

---

[4] Technical Datasheet DS56, Power Light Source Luxeon Rebel, Philips Lumileds Lighting Co., 2007.

**TABLE 8.1 Efficacy of LED at 350 mA, Including Temperature Effects**

| Parameter | Value | Units |
|---|---|---|
| I | 350 | mA |
| Vf | 3.15 | V |
| Power dissipation | 1.103 | W |
| Pad temperature | 36.5 | °C |
| Temperature efficacy | 96 | % |
| Efficacy | 57.6 | Lm/W |
| Light | 63.5 | Lumens |

500 mA it is approximately 1.30. The efficacy has thus decreased from approximately 1.00/0.35 = 2.86 to approximately 1.30/0.5 = 2.60, a 9% decrease.

Putting these two mechanisms together will result in a higher power required to get the same light output with ripple current. This leads to a higher temperature, shortening lifetime of the LEDs. There is a second-order effect as well. The increased temperature also decreases the efficacy. But for practical purposes, this effect is usually small enough to be ignored.

Let's look at a practical example to illustrate the effect. The numbers are summarized in Tables 8.1 and 8.2. As a baseline, let's take a Luxeon Rebel being operated at a DC current of 350 mA, with a thermal resistance of 15 °C/W to an ambient of 20 °C. Suppose it has an output of 60 lumens at 20 °C.

Now at 350 mA, the forward voltage of the LED is 3.15 V, so the power dissipation is 3.15 V·0.35 A = 1.103 W. This gives a device temperature of 20°C + 1.103 W·15 °C/W = 36.5 °C. At this temperature, the efficacy is reduced to 96%, which is 60 Lm/W·0.96 = 57.6 Lm/W. Light output is thus 57.6 Lm/W·1.103 W = 63.5 lumens.

Now let's examine one with a ripple current. We still want an average of 350 mA, but superimposed on top of this is 200 mA peak-to-peak ripple (see Fig. 8.24). Peak current is thus 450 mA, and minimum current is 250 mA. Here's where the practical part comes in. Rather than trying to integrate the various parameters over the triangle wave, we're going to approximate the current as constant at 250 mA for half the cycle and at 450 mA for the other half.

**TABLE 8.2 The Efficacy at an Average Current Is Less Than That at That DC Current**

| Parameter | Value at 250 mA | Value at 450 mA | Average | Units |
|---|---|---|---|---|
| I | 250 | 450 | 350 | mA |
| Vf | 3.10 | 3.25 | — | V |
| Power dissipation | 0.775 | 1.463 | 1.119 | W |
| Pad temperature | 36.5 | 36.8 | 36.8 | °C |
| Temperature efficacy | 96 | 96 | 96 | % |
| Relative light output due to current | 0.76 | 1.23 | — | — |
| Efficacy | 54.2 | 48.4 | 51.3 | Lm/W |
| Light | 42.0 | 70.8 | 56.4 | Lumens |

Figure 8.24    200 mApp ripple current on a 350 mA DC drive.

Now at 450 mA, the forward voltage increases from 3.15 to 3.25 V. At 250 mA, the forward voltage decreases from 3.15 to 3.10 V. Power dissipated during the 250 mA part of the cycle is therefore 3.10 V·250 mA = 0.775 W. During the 450 mA part of the cycle, it is 3.25 V·450 mA = 1.463 W. Note that 2/3 of the power is dissipated during the high part of the current cycle. But average power is essentially identical with the DC case, as was to be expected. Thus, device temperature is also almost identical, now 20°C + 1.119 W·15°C/W = 36.8°C. Since the temperature is the same, the temperature effect on the efficacy is the same, 96%. However, efficacy is also affected by the current level. At 250 mA, 76% of the light at 350 mA is produced, and at 450 mA 123% light comes out. Calculating efficacy at 250 mA, 60 Lm/W· (0.76/0.775 W)·96% = 54.2 Lm/W.   At   450 mA,   60 Lm/W·(1.23/1.463 W)·96% = 48.4 Lm/W. Light emitted during each condition would be 54.2 Lm/W·0.775 W = 42.0 lumens and 48.4 Lm/W·1.463 W = 70.8 lumens. Total light output is the average of the two, 56.4 lumens.

Light output has therefore decreased 11% due to the ripple current. Looked at the opposite way, to get the same light output would require 1/(1 − 11%) = 12% more power—which would increase the losses even more. The conclusion is that ripple current can have a significant impact on the light output from your LEDs. If you want to get it right the first time, you have to account for ripple—or reduce it to a level where it's insignificant for the application.

## LIFETIME

There are some additional factors affecting lifetime of your design, beyond those already mentioned in the chapter on DC drivers. These additional factors are 60 Hz ripple in the input capacitor, lightning, and ripple current in the LEDs.

Let's start with the input capacitor. In Chapter 7, we noted the need for a high-frequency capacitor at the input of switch-mode power supplies to provide low impedance for the high-frequency power pulses. This is still true in an off-line converter. However, the input capacitor in AC converters in addition has 60 Hz (or 50 Hz) ripple. And it turns out that this is often the dominant ripple source. When you look at the AC current in the capacitor with a current probe, most of what you see is at line frequency.

There's really nothing new here, except that the capacitor ESR again has to be selected to ensure that you don't degrade the MTBF of the design. And off-line converters sometimes use large electrolytic capacitors. It is these that have limited lifetime. As discussed in previous chapters, there's usually not enough data on a capacitor datasheet to tell what the ESR is. So the usual method is to build the converter and measure the current. This is helpful for ensuring you're not exceeding absolute maximum ratings. But from a practical standpoint, lifetime is set by the capacitor temperature. So again, sticking a thermocouple on the capacitor tells you the lifetime. For most applications, it's possible to find a capacitor with a high enough lifetime at a high enough temperature to meet the MTBF. The problems usually arise when it also has to be physically small enough. The solution is to power factor correct the SMPS, so no electrolytic capacitor is needed.

There are frequent comments in the literature about limiting the ripple current to the LEDs. It is claimed that ripple current degrades the lifetime of the LEDs, but few specifics are given. It turns out that for reasonable levels of ripple, the dominant effect on LED lifetime is the increase in dissipated power, as we've seen above. This means that the normal calculation and measurement of LED temperature is enough to ensure their lifetime. But what is a reasonable level of ripple? The answer is that the LEDs' peak current should not exceed the absolute maximum rating of the device. As long as this is respected, the LEDs' lifetime will be reasonably well predicted by their temperature.

The final new factor affecting lifetime of the system is lightning. We've already seen that the best you can do about lightning is to put in the biggest MOV you can fit. But is that good enough? Because MOV clamp voltages degrade with usage in a nonspecified way, there's no analytical answer possible. Here's a practical approach to the answer. Following the guidelines given above for component selection, the voltage let-through by the MOV will not be enough to damage the converter. Thus, lightning won't cause the unit to fail until the MOV fails. How long does this take? We see units in the field that are not very well protected (say, with a 7 or 10 mm MOV) that last for 10 years or more. While the frequency of lightning depends on the geographic area in which the unit is installed, in general a 10–14 mm MOV should be sufficient for most locations. Of course, if you can afford the space for a 20 mm MOV, that is preferable.

## UL, ENERGY STAR, AND ALL THAT

In this final section, we're going to survey some of the governmental and quasi-governmental regulations surrounding LED lamps. As noted in Chapter 2, governments have seen fit to regulate lighting, and there is a wide variety of safety and other agency requirements as well (see Table 8.3) (Dowling, 2009).

Indeed, some would say that governments have seen fit to overregulate LED lighting. A prime example of this is the U.S. government's specification of power factor for LED bulbs. In the United States, the Department of Energy awards an Energy Star rating to devices that are in the top 10% at saving energy. To the extent that consumers look for the Energy Star insignia, this has helped to increase the average efficiency of items such as refrigerators and water heaters. These are all goods with well-established markets. With LED bulbs, however, the DOE decided to take the initiative, and put out requirements that LED bulbs need to meet Energy Star *before* the market was established. How do they know what the top 10% of bulbs do? For example, Energy Star LED bulbs will be required to have a minimum power factor of 0.7 (residential) or 0.9 (commercial). Adding power factor correction to an LED ballast costs money. In some cases, it may make it impossible to make the bulb

**TABLE 8.3  A List of Standards**

Photometry
    CIE 127-2007 (TC2-45) Measurement of LEDs
    IESNA LM-79 Electrical and Photometric Measurements of SSL Products
    IESNA LM-80 Method for Measuring Lumen Maintenance of LED Light Sources
    CIE TC2-46 CIE/ISO LED Intensity Measurements
    CIE TC2-50 Optical Properties of LED Arrays
    CIE TC2-58 Luminance and Radiance of LEDs
    IESNA TM-21 Predicting Lumen Maintenance of LED Sources

Color
    ANSI C78.377-2008 Chromaticity of SSL Products
    CIE 177-2007 (TC1-62) Color Rendering of White LED Light Sources
    CIE TC1-69 Color Quality Scale (new CRI)

Photobiological Safety
    IES RP-27 Photobiological Safety
    IEC 60825-1-2001 Safety of Laser Products (to be Superseded)
    CIE S009 Photobiological Safety
    IEEE P1789 Recommended Practices of Modulating Current in High Brightness LEDs for Mitigating Health Risks to Viewers

Safety
    ANSI C82.SSl1 Power Supply
    ANSI C82.77-2002 Harmonic Emission Limits
    ANSI C78.09 82 Fixture Safety Specification
    FCC 47 CFR Part 15 Radio Frequency Devices
    IEC SC 34A 62031:2008 LED Modules – Safety
    IEC SC 34C 61347-2-13:2006—Lamp Control Gear Part 2-13: DC or AC Control Gear for LED Modules
    IEC SC 34A IEC 62560 Self-Ballasted LED Lamps
    IEC SC 34A [TBD] LED Lamps >50 V—Safety Specs
    UL 8750 LED Light Sources for Use in Lighting Products

Performance
    IEC SC 34C 62384-DC or AC Supplied Electronic Control Gear for LED Modules
    EC SC 34A-Performance Standard for LED Lamps

Nomenclature
    IES RP-16 Nomenclature and Definitions Addendum A: SSL Definitions
    IEC SC 34A-TS 62504 Terms and Definitions for LEDs and LED Modules

EMC and Other
    IEC TC 34 EN 62547 LED EMC/Immunity
    IEC SC77A-EN 61000-3-2 LED EMC/Harmonics
    ANSI SSL2 LSD-45 Sockets & Interconnects
    ANSI C82.04 Driver Safety Circuitry

*Source:* Dowling (2009).

(think of fitting the extra circuitry into an MR16 form factor). The government is thus mandating performance, without regard to the effect of this mandate on the market.

There are many other potentially onerous requirements for Energy Star. As a minor example, LED bulbs are required to meet FCC EMI requirements of Part 15.

While it seems reasonable that all electronic devices should do so, in fact CFLs are required to meet only Part 18, which is somewhat easier. Why should LED lighting be singled out for different treatment? Our guess is that LEDs have been touted by enthusiasts as the perfect light source. Since they are (or rather will be the enthusiasts claim) perfect sources, the government has decided to require them to be better in every way, regardless of whether or not the market is willing to pay for these improvements. It remains to be seen whether this hurts the market for LED bulbs.

UL also has quite a number of standards that will potentially affect LED lighting (see Table 8.4). Of course, the venerable UL1993 is still applicable. But they have now added UL8750, "Safety Standard for Light-Emitting Diode (LED) Equipment for Use in Lighting Products." The most significant requirement, as always for AC-powered equipment, is line isolation. A consumer handling the bulb in a reasonable way should not be able to come into contact with the AC line. But what is reasonable? Clearly, having live wires hanging out is too dangerous.

An easy solution is to make the ballast isolated. But this involves (at least) changing from an inductor to a transformer. This certainly adds to the cost, and probably the size and weight of the ballast. Are other solutions possible? While each UL evaluation is individualized, apparently bulbs in which the live electronics (and traces, solder pads, and other components) are covered with a plastic or glass cover are reasonably safe. Recently, individual LEDs have been recognized as being safe. After all, even if all of the electronics are covered up, the LEDs have internal bond wires that

**TABLE 8.4  UL Standards Affecting LED Lighting**

| Standard | Product type |
| --- | --- |
| Fixed luminaires | |
| UL 1598 & CSA C22.2 #250 | Luminaires |
| UL 1573 | Stage and studio lighting |
| UL 1574 | Track lighting |
| Portable luminaires | |
| UL 153 | Portable electric lamps |
| UL 1993 | Self-ballasted lamps and lamp adapters |
| Specialty luminaires | |
| UL 48 | Electric signs |
| UL 676 | Underwater lighting fixtures |
| UL 844 | Fixtures for use in hazardous locations |
| UL 924 | Emergency lighting and power equipment |
| UL 1786 | Nightlights |
| UL 1838 | Low-voltage landscape lighting systems |
| UL 1994 | Low-level path marking and lighting systems |
| UL 2108 | Low-voltage lighting systems |
| UL 2388 | Flexible Lighting Products |
| Power supplies | |
| UL 1012 | Power units other than Class 2 |
| UL 1310 | Class 2 power units |

*Source:* Straka (2009).

could be connected to AC line. No general formula can be given for passing UL, although a safe route is to use an isolated ballast. There are ways around this, and you have to talk to UL about them (preferably before getting very far into the design).

One other area of safety regulation that is almost unique to LED lighting is eye safety. There is a long history of limits on lasers and how it can be ensured that people aren't blinded by them. But with individual LEDs getting brighter and brighter, and with the limited light emission angle some devices give, LEDs are a potential threat to eyes as well. There are now some limits in place on how bright an individual device can be, without requiring some sort of diffuser or other means to limit the brightness. For example, there is an optical safety requirement in the EU's EN60825-1. Although not an immediate problem for LED lighting, this is something to keep in mind for the future.

CHAPTER **9**

# PRACTICAL SYSTEM DESIGN WITH LEDs

IN PREVIOUS chapters, we have examined the components that make up a lighting design: the LEDs, the driver, their thermal performance, and so on. To make a complete design, however, these various components need to be integrated. We need to make a printed circuit board for the LEDs and for the ballast. We need to design optics to get the light from the LEDs to go where we want it to go and to have the characteristics we want. We need the design to survive in the customer's environment. And we also need to look forward to our next design, when the efficacy of the LEDs will have increased yet again. It would be wise to make our design flexible enough to accommodate this next generation of components without starting the design all over again. After all, those better components will be here in just a few months.

## PCB DESIGN

There are specialists who do nothing but lay out printed circuit boards (PCBs.) And within this group of specialists are specialists who focus on power supply layouts. Given this degree of specialization, we won't be able to tell you in detail how to lay out PCBs for your LEDs and ballast. What we're going to do is to give some pointers on concepts we have found to work well based on long years of experience. We'll also advise on pitfalls to avoid, based on PCBs that came back and didn't work quite right.

The first step in creating a good PCB is to make a good schematic. A good schematic is more important than it at first seems, for two reasons. In the first place, a good schematic is easy to look at and understand. This is important for anyone else who has to look at the schematic, for example, for a consultant or for a design review. It's also important for you, when you have to go back and look at that old design 12 months from now.

A second reason a good schematic is important is because it helps generate a good layout. When you spend enough time to lay out a good schematic, it becomes a

*Practical Lighting Design with LEDs*, Second Edition. Ron Lenk and Carol Lenk.
© 2017 by The Institute of Electrical and Electronics Engineers, Inc. Published 2017 by John Wiley & Sons, Inc.

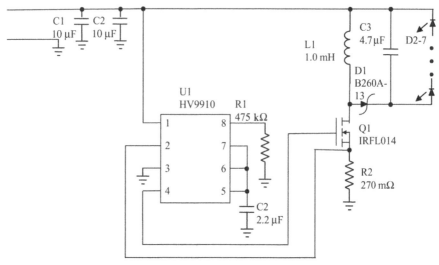

Figure 9.1  A good schematic.

guide telling you where components should go on the PCB. Components that are "naturally" close together on the schematic often should be close together in the layout as well.

Look at Figure 9.1 for a sample of a good schematic versus Figure 9.2 for a schematic that could be improved.[1] The most immediately obvious difference is in the IC pin-out. The original schematic shows the pins out of numerical order and with some pins on all four sides. Probably this is to minimize the visual complexity of the schematic. But it's not useful for laying out a board because the IC doesn't come with pins that do that. The improved schematic shows the IC pins in their actual order. This makes it obvious that the IC connections to the gate and source of the MOSFET are on the same side as $V_{IN}$ and GND. The pins thus have to either be routed around the IC (possibly a noise problem with the gate drive) or the IC could be rotated 180° or be mounted on the backside of the PCB. Conscientiously putting the pins in the right place on the schematic will make it obvious if connections need to cross over other connections, a frequent occurrence with ICs. This can be an EMI problem or even a routing problem if you're trying to make a one-layer board.

The next aspect to notice about the improved schematic is the reference numbering. The original shows capacitors with reference designators C4, and C6 through C8. Where are C1 through C3, and C5? Reference designators should start from the number one and proceed sequentially without skipping numbers. Otherwise, the BOM will be confusing, and buyers and others will ask questions about where the missing parts are.[2]

---

[1] Don't get us wrong, we think the 9910 is the best high-voltage LED driver on the market today. It's just the schematic that could use some improvement.

[2] If you have two identical sections that you want to reference the same way, you could call the parts in one section 101, 102, and so on, and the corresponding parts in the second section 201, 202, etc.

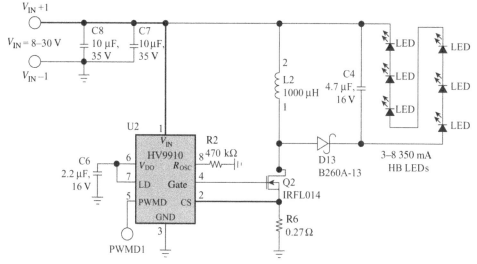

Figure 9.2   A schematic that could be improved. (*Source:* HV9910 datasheet, Supertex Inc., 2006.)

A further reference designator problem is that the references of the original schematic have no order. C8 is to the left of C7, and C4 is way off to the right. If you have a hundred capacitors in the schematic, how can you find C47? We recommend that the proper way to draw the schematic is to start numbering components from the upper left, and proceed toward the lower right. Of course, there will be some components whose sequential numbering will be not as clear, but overall this makes finding a given part much easier.

There are still improvements that could be made to the improved schematic. These would be more important for a design with more components. For example, you could show the ground for power components separately from the small-signal component ground. They could use different ground symbols, or have a "p" and an "s" attached to the ground symbol, respectively. In the current schematic, small-signal ground would be only for R1 and C2, and the IC ground would be where the two grounds connect.

In some cases, it may also be useful to color-code the schematic. For power supply schematics, we find it useful to indicate the high-current traces in one color (say, yellow) and the high-voltage traces in a second color (say, red). That way the PCB layout person knows that yellow traces have to be wide and red traces need to be well separated from other traces. The former is for current-carrying capacity, the latter for arcing. You might also consider another color for signals with high frequencies, to be careful with EMI and cross talk.

One ambiguous area of the schematic is the placement of R2. It measures the current from Q1, and it seems natural to put it there. On the other hand, the actual layout will have R2 right next to pins 2 and 3 of U1, to minimize noise in the feedback path. Which one should you pick for your schematic? Since there's no good solution, you just pick one. Maybe you can add a note about placement to the schematic.

Here is a checklist of design elements to aim for when putting together a schematic:

1. Draw the schematic with real pin-outs.
2. Order reference designators starting from 1, and don't skip numbers.
3. Assign reference designators, starting from the upper left-hand corner and working toward the lower right-hand corner.
4. Have separate symbols for signal and power grounds.
5. Color code high-current, high-voltage, and high-frequency traces.
6. Make notes about components that should have special attention during layout.

Now we have an easy-to-use schematic. Our PCB is basically going to look like our schematic. We place the components approximately in the relative positions they occupy on the schematic, and the connections between them route approximately the same as they do on the schematic. There are a few additional considerations. Let's talk about grounding first.

## Grounding

Your PCB should have a good ground system. Just connecting all of the grounded components together with traces really doesn't work. What you want is to connect all of the power grounds together with thick traces or a plane, a copper pour. Power grounds are all the grounds carrying substantial currents, say more than 100 mA. Use as much copper as space permits. Once you've connected all of these grounds together, it should connect back to the input capacitor's ground.

Don't make loops in the power ground. A loop means that there are two ways for the current to get back to the input. This creates an EMI problem. Of course, a via in the middle of the ground plane is acceptable, and usually necessary. Just ensure that the via isn't right at the edge of the plane, which would give a small trace on one side of it. It's better to retract the plane a little away to avoid having the small trace.

It's best to let each ground go directly back to the input via the plane, rather than trying to connect one trace to the next in a chain. All power grounds should go directly to the ground plane, directly if possible and otherwise through one or more vias.

The small-signal grounds should also be attached together using traces or a separate small plane. Don't just hook them into the power ground. This separate plane is called an island. Its purpose is to keep the high currents on power grounds from causing voltages on sensitive nodes. For example, when the MOSFET turns on, there's a fast current spike from its source to ground. Since even ground planes have resistance and inductance, this can cause a fast voltage spike on the ground. This spike may be large enough to cause false triggering of the IC. Connect the island to the power ground plane at a single point. We recommend the connection to be at the input capacitor, if possible. Multiple connections back to the power ground generate the noise you're trying to avoid.

Avoid chains if possible, make the signal ground into a star. A chain is a series connection of component grounds. For example, you have two capacitors and two resistors all of which are grounded, and you attach a trace from the small-

(a)

(b)

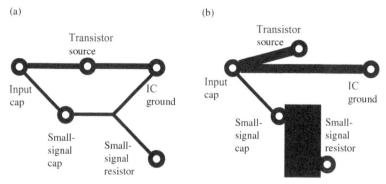

Figure 9.3    Poor grounding layout (a) and improved layout (b).

signal ground to each of them in turn. The one at the end of the chain will be affected by the noise from all of the others. A star gives each component its own trace back to the ground, so that all small-signal grounds see the same noise simultaneously.

Figure 9.3 gives some idea of what should be avoided (a) and what should be done (b). On the left, the ground of the IC is returned to the input capacitor's ground by going through the transistor's source connection. The high-frequency components of the two will mix, potentially causing the transistor—or the IC—to turn on or off at the wrong time. On the right, the IC ground and the transistor's source are returned to capacitor ground separately. On the left, the IC ground has *two* paths back to capacitor ground: one through the source and another through a path through a small-signal ground. This loop splits the current from the IC ground, creating a ground loop that can affect EMI. On the right, the IC ground has a single path back to capacitor ground. On the left, the small-signal capacitor and resistor are in line with the current from the IC ground. Since the resistor joins this path before the capacitor does, they will see different ground voltages. They will also see different noise. This can potentially upset the operation of the IC, resulting in things such as missed switching cycles. On the right, the small-signal components' grounds are tied together on an island, and then returned by their own separate trace to capacitor ground.

## Other Connections

Power connections should be treated much as power ground is. For example, the connection of the MOSFET drain, the inductor, and the anode of the diode in the schematic of Figure 9.2 should be connected with a plane to the extent possible. Wide traces are acceptable, but make them as wide as will fit. Extra copper is free!

What constitutes a high-voltage trace is somewhat relative, but generally anything above 60 V is appropriate. There aren't any on our schematic, although the drain will come close. You should leave extra spacing between this type of traces and others at low voltage, to avoid any possible arcing problems.

High-current traces, including power traces, should be wider than normal. In general, the wider the trace, the better. This minimizes voltage drops and efficiency losses. On a related topic, we recommend that all ballasts and all PCBs with LEDs should use at least 1oz copper, not the standard ½oz. It is even better to use 2 oz copper, if the cost is not prohibitive.

High-frequency traces need to avoid crossing other traces. Ideally, of course, they should be very short. But if it's necessary to run it some distance, running it above the ground plane is good. This minimizes the loop area for radiation. The gate drive from the IC to the MOSFET is particularly a problem. It has both high current and high frequency. This trace needs to be short, if possible. It is probably better to move the MOSFET than to have a long trace for this signal.

The following is a checklist of things to aim for in laying out a PCB:

1. Make ground a plane or a portion of a plane. Connect all of the high current grounds directly to it. Avoid loops and traces.
2. Make a separate ground island for small-signal grounds. Connect it at a single point to power ground.
3. Connect power connections with a plane. If you need to use traces, make them as wide as possible. The same is true for all high-current connections.
4. Separate high-voltage traces from others with extra space.
5. Use at least 1oz copper for the PCB.
6. Keep high-frequency traces short, and don't cross other traces with it. Avoid vias. The gate drive in particular needs to be short.

## GETTING THE LIGHT OUT

The purpose of an LED light is to illuminate. The LEDs produce the light, but there is usually more to it than that. You want the light to have a certain distribution, bright in some places and dark in others. For example, a spotlight needs to have most of its light within a certain forward angle. You want the light to be of a specific color. For example, daylight incandescent bulbs use the same filament as normal incandescent bulbs do, but modify the light color by adding neodymium to the glass. And you want the light to be bright enough. If you've spent all that money on LEDs, you don't want to waste 30% of it in a piece of yellowing plastic.

All of these aspects are part of the system design of the optics. We'll start with lenses. Everyday experience gives us the idea of what lenses can do. In the form of a magnifying glass, they can bend and focus light. As a prism, they can split white light up into a rainbow of its constituent colors. A colored lens can absorb light of a particular color, removing it from the incoming light. And, while less obvious in daily experience, lenses also reflect light to some degree.

While lenses are used for some of these things in LED lights, some of the other things mentioned above we try to avoid. For example, the primary optics in an LED can be used to direct edge-emitted light from the LED forward, so that light comes out of the front rather from the sides of the device. The silicone covering of an LED also acts as a lens and helps to reduce light loss. Since the index of refraction of the die and air are quite different, the silicone acts as an intermediary. By lowering the index of refraction in two steps rather than doing it all at once, reflections from the interface can be reduced. This increases the light output from an LED.

Of course, these lenses also have some undesirable characteristics. The primary optics in an LED don't redirect all of the light. Some of it is absorbed, some of it is directed in the wrong direction, and some just reflects back to the emitter and is lost as heat. The die-to-silicone interface and the silicon-to-air interface have similar problems. Additionally, silicone eventually yellows. This absorbs part of the light spectrum, reducing efficacy and changing the color of the emitted light.

Beyond lenses on the LEDs themselves, there are other optical devices, whether their optical properties are intentional or not. A flashlight may have a glass or plastic lens in front of the LED in order to obtain the right beam angle. A USB light may have a glass or plastic cover so that the user is not accidentally burned by the hot LED. And an LED light bulb may need to have a cover over it to avoid electrical shock hazard. It will probably also have overall structural elements to make it work mechanically and aesthetically, and these can reflect or absorb light as well.

As suggested by these examples, glass and plastic frequently form part of the interface between the LED light and the outside world. Looking first at glass, we note that dealing with glass is almost a forgotten art in the world of electronics. It takes some searching to find someone who knows enough about glass to make a design that is feasible with the material, mechanically robust, and cost-effective. To start with, there are a number of different common types of glass: soda-lime glass, borosilicate glass, and lead glass, among others. Soda-lime is the most common and least expensive type, used for bottles and jars. Lead glass is used in stemware and in incandescent light bulbs (it really does contain lead). Borosilicate glass is the most expensive, used in chemical test tubes and the like. It has great resistance to thermal shock, and low coefficient of thermal expansion.

In practice, your lights, if they use glass at all, will probably use soda-lime glass. Soda-lime glass has relatively good optical transmission in the visible region (see Fig. 9.4). It's in the mid-90% range near the photopic peak in the green. It also transmits well throughout the IR, so that heat can escape by radiation.

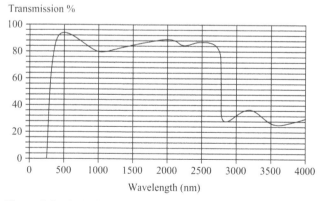

Figure 9.4  Soda-lime glass optical transmission. (*Source:* http://en.wikipedia.org/wiki/File :Soda_Lime.jpg under license http://creativecommons.org/licenses/by-sa/3.0/.)

Now while 92–94% transmission is very good for an optical device, as soon as you add glass to the optical path from the LEDs to the world, you've lost 6–8% of the light you struggled so hard to make.

Glass also has the downside of being easy to break. Maybe this isn't a significant problem; incandescent bulbs have been made from glass for a century, and fluorescent tubes are made from glass and even contain a neurotoxin. On the other hand, UL thinks that glass is decorative, and doesn't consider it to be a barrier against AC line voltage. Thus, it may be a marketing decision as to whether plastic rather than glass is used.

Plastic doesn't usually shatter like glass, but it presents other concerns. For one thing, there are many types of plastics, such as polyethylene, polyethylene terephthalate (PET), and polycarbonate. And they all have different properties, which will probably matter to your design. There are also many different grades of each type. How are you going to pick the right kind? The easy way is to pick the same kind that the competitors are already using. They must know what they're doing, right?

The first thing to know about plastics is that they are typically less optically transmissive than glass. Glass will lose 6–8% of the light that strikes it. Even the best plastics are more like 10%, and some are much higher than that. So you're going to have to put more power into the LEDs to obtain the same light output. You could pick a plastic with really good clarity, such as polycarbonate, but the cost of this is higher than other choices.

Plastics also have temperature limitations. Glass of course can be exposed to high temperatures without effect. But plastics can soften at very low temperatures, and completely melt not much above those. For example, polyethylene melts between 105 and 130 °C, depending on what type you use. PET starts softening at 75 °C. Some polycarbonates, on the other hand, don't melt until 265 °C, and can be used well over 150 °C. In fact, some plastics will burn if put into a fire, and may release toxic gases. What will firefighters think about the presence of such a plastic in a burning house that contained your light bulb?

Plastics also yellow with age. Both heat and light are accelerators for this type of aging—naturally, just the characteristics you expect to see in an LED light. Of course, some plastics are better than others. In particular, polycarbonate is used in automobile headlights. They can sit in the sun for 10 years and be right next to high-temperature incandescent bulbs without noticeable yellowing or hazing.

Some plastics diffuse light. Sometimes this is intentional, as when you want the light to be spread out more uniformly than when there are only a few point sources of light. The diffusion, however, is not loss-free. Diffusion results in increased optical loss, from 10 to 30%. So you have to crank up the LEDs even more to compensate. We should mention that the soft-white-type incandescent bulb has very low loss in its diffusion. We have been told that it is only a few percent, so perhaps this would be a good model for manufacturers of LED bulbs to copy.

The conclusion is that almost anything you do to the emitted light is going to cause optical loss. If possible, it's better to get LEDs that have the right emission direction than to try to shape the light with lenses. Glass covers have optical losses. If the lack of mechanical robustness of glass is a problem, you could go with plastics. As you can tell, we prefer polycarbonate. But plastics absorb more light than glass does, and adding diffusion absorbs even more light. It's probably best to get a consultant who knows all about plastics. (It's a good thing that higher efficacy LEDs are coming out next month!)

# LEDs IN HARSH ENVIRONMENTS

Most of the study we've done until now has been for LEDs operating in a lab bench sort of environment. They get hot when you put power into them, but the environment is generally benign, 25 °C or maybe 40 °C at worst. The same assumption has been true for ballasts. But in a real system, the LED and the ballast heat each other, the mechanical part of the system may affect the LEDs, and the environment may not only be hot but unfriendly to LEDs and circuitry in a variety of other ways. For example, would LEDs work properly in a microwave oven? The authors don't know and would be glad to hear from someone with experience.

Some environments are going to be far hotter than others. An LED in a system left sitting on a car dashboard in summer in Florida will see an ambient temperature of 50 °C. An LED in a lighting system in a traffic light in Alaska will see −40 °C or lower. LEDs are unaffected by either of these extremes (although the electrolytic capacitors in the ballast won't work at −40 °C). But they might be affected being in a display window in Alaska that is turned on only from 4 to 9 p.m. each night. They might be up at 85 °C when they're on, and then drop to −40 °C when they're turned off. Now they are being put through a wide ranging thermal cycle every day, 365 times a year. Thermal cycling tends to break things, because almost everything physically expands when hot and contracts when cold. Solder joints loosen, plastic cracks, bond wires can fail. Any of these can potentially cause your light system to fail. In military designs, thermal cycle testing of the design is a requirement. It may also be a good idea for your design.

Another problem is also caused by thermal cycling that goes through 0 °C. When a system is cooled below 0 °C, humidity in the air tends to become frost. Then when it warms up again, water droplets may form. On a circuit board, these can cause electrical short circuits, and thus failure. On a ballast, it's common practice to seal the PCB with a conformal coat to avoid this problem. But on a PCB for LEDs, this may not work. What are the optical losses in the conformal coat? Will the conformal coat interact with the silicone on the LED?

Another potential environmental hazard for LED lighting systems is humidity. Humidity can degrade phosphor performance, leading to premature dimming. Companies routinely ship LED reels in sealed bags with silica absorbent to prevent moisture from compromising the parts. Some recommend baking LEDs from opened bags for 24 h to evaporate the moisture. The fear is that when the LED is soldered, the moisture will form steam, potentially damaging the part. In any case, the effect of humidity on your system should be experimentally evaluated. In Malaysia, for example, it can be close to 100% humidity almost all the time.

A related problem is salt fog. In places close to the ocean, houses need new coats of paint every 2 years. This is because it is always humid there, and the air contains lots of salt. Salt corrodes paint and metal, among other things. So if, for example, you're making a light that can be used outdoors, you need to check that a salty, humid environment won't cause your light to fail prematurely.

These problems with temperature and humidity are exacerbated if your system contains plastic, for example, as the external case of your light. As we've said, there are many different kinds of plastic. Many aren't very good at high temperatures. They can become soft and lose mechanical strength, or they can become brittle and shatter easily. Many plastics will turn yellow with heat (or sunlight) and this may affect the color of the light that your system puts out. Again, if you plan to use a plastic in your product, you should turn to a plastics expert to ensure you're doing it right.

LEDs can be blown up by ESD. But a lighting system can also be affected. You have to ensure that the customer can't destroy it by touching the case or the electrical inputs with a static charge. Systems attached to the AC line are worse. Although we've shown protection for the ballast by adding in an MOV, *some* voltage always gets through. MOVs increase in voltage depending on how much current passes through them. If there's a lightning strike close enough to the light, it will be destroyed no matter what you do. Also remember that MOVs degrade with number of surges. If you have line surges 10 times a day from a nearby motor starting up, this may damage the MOV enough that it no longer provides sufficient protection against lightning. The only practical solution is to put something that doesn't cost too much on the market, and evaluate returns to see if it is cost-effective to increase the protection versus unit replacement and dissatisfied customers.

By the way, we'd like to offer this unpleasant thought. Incandescent bulbs are like resistors. If lightning comes along, the incandescent bulb absorbs a good deal of the energy. Of course, a direct hit will destroy anything, but the incandescent bulbs are suppressing some of the energy of some of the neighborhood's lightning. What happens when all those incandescent lightning protectors have been replaced with LED lights that are just electronics?

LEDs are semiconductors, and as such can be affected by vibration. When you put an LED light on an automobile, it experiences many bumps and jolts over its lifetime. Bond wires can fail in the LED from these vibrations, the same as they can fail in the controller in the ballast. Fortunately, in order to evaluate them, standardized tests have been developed for this sort of environment.

Think all those environments weren't enough? Lights can also be affected by dust. Dust settles on everything. Light bulbs are rarely cleaned, if ever, and thus the light output from the bulb dims with time, even if the light source itself stays constant. A device like an LED light bulb can collect even more dust than furniture; the presence of a high voltage on the surface, even though it is safe and passes safety regulations, may actively attract dust. Think of how the monitor on your laptop computer attracts dust. Although companies claim that this dust depreciation is "not our fault," the customer will still perceive that your bulbs have become dimmer with age, without thinking about why.

## DESIGNING WITH THE NEXT-GENERATION LED IN MIND

Way back in Chapter 1, we learned the benefit of Haitz's law. Every few months a brighter and cheaper LED comes out, and all of our lighting products become brighter and less expensive just by changing part numbers. Indeed, you can simply plan on this being true. But this comes with a downside too.

For one thing, probably we've designed with enough LEDs to obtain the required brightness in the first place. If we now substitute brighter LEDs, perhaps the light will be *too* bright. The customer wanted 630 lumens, not 750. In a traditional engineering environment, this would be viewed as a cost-cutting opportunity. We can now make a new design that uses fewer LEDs. We can reduce the cost of the heat sink, and the ballast can be redesigned to produce less power. Unfortunately, all of this takes time and engineering resources. Redesign of the ballast can take a month, and laying out new PCBs for the ballast and the LEDs can also take a couple of weeks. The analyses have to be redone. All the tests have to be run again: thermal, EMI, and aging. Maybe UL qualification has to be redone. But before this new design is implemented, the *next* new LEDs will be out!

While this may sound like assured employment for the engineering department, the reality is different. Management may decide to skip a generation to get adequate return on the engineering expenses already incurred, or you may have inventory that has to be used up before the next design can be produced. And then your competitors will come out with a better bulb while you're in between versions. The situation is analogous to the microprocessor world, where some years ago a faster clock speed came out every month. Your designs simply have to plan on that.

The right way to deal with these problems is to plan for obsolescence. If your design needs 18 LEDs, it isn't too hard to figure out that in the near future it will only need 16, and not too far down the road it will only need 12. So when you design the ballast, verify that it works not just with the 18 LEDs you're using today, but also with the 12 you'll be using six months from now. That way you don't have to redesign and requalify the ballast, and you will save time and money.

Your LED PCB can be laid out to accommodate different numbers of LEDs. If you need the LEDs to be mechanically symmetrical, you could include extra footprints on the PCB so that it remains symmetric when you go to fewer devices. The extra footprints can be shorted out with a jumper when you don't need them. For that matter, the PCB can be laid out so that each device has multiple footprints, enabling you to switch LED vendors without laying out the board a second time.

## LIGHTING CONTROL

One other aspect of lighting system design is the possibility of an interface of the light with a lighting controller. None of the designs in this book cover this possibility, because it is a specialty item and takes substantial time and money to implement in both hardware and software. Nonetheless, there may be circumstances in which you need to communicate with your light. As a particular example, large installations of lights in office buildings may need to be under central control. For example, they may all need to be turned off at 8 p.m. when everyone has gone home, and on again in the morning, so that a person doesn't have to physically walk around every floor turning them on and off every day.

We admit to not being experts in the field of lighting control. So here we're going to take some extracts from Wong and Zheng (2009), and leave it at that.

## DALI Protocol

"Digital dimming using the DALI protocol would add more flexibility and the advantages of being able to individually control up to 64 luminaires from a single pair of signal cables. This approach offers significant advantages in larger installations such as offices, particularly where intelligent lighting systems are used that employ advanced energy management techniques such as presence detection and daylight harvesting.

"DALI is a digital addressable lighting interface designed to replace and enhance the traditional 0–10 V analog dimmer. It is an open-lighting control protocol backed by major lighting manufacturers, and it has been established as the *de facto* control standard in fluorescent ballasts.

"The two-wire DALI network supports up to 64 lighting devices. Since Manchester encoding is utilized, the polarity of the wires is interchangeable, resulting in fool-proof installation. Since the topology of the devices can be a star, a bus, or in a daisy chain, retrofitting is simple when expanding a network (see Fig. 9.5). The communication distance is less than 1000 ft and is based on a master–slave arrangement, the master being the DALI controller and the slave being the ballast. A bidirectional communication enables the ballast to return its condition and setting back to the DALI controller. The DALI protocol has a relatively low baud rate of 1200 bits per second (bps), but is sufficient to turn on, off, and dim a light. Since DALI is an addressable protocol, devices can be addressed individually or by group. A DALI master device can control a single light or a group of lights, and DALI command has been extended from typical level dimming and scene setting to incorporating environmental sensors for ambient light and motion sensing. The DALI standard also includes extensibility of manufacturer-specific commands."

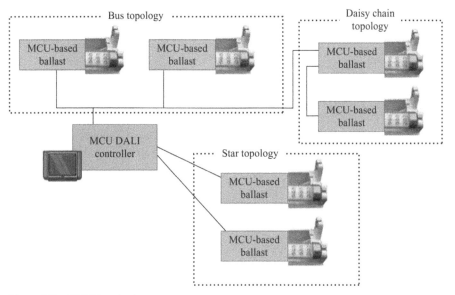

Figure 9.5   DALI topologies.

## DMX512 Protocol

"DMX512 is a digital multiplexing lighting protocol that is popular in theatrical, architectural and entertainment lighting and can control up to 512 devices."

## 802.15.4 and ZigBee Open-Standard Technology

"The ZigBee protocol has been spearheaded by utility companies to enable smart energy inside a home, where different appliances can communicate and control themselves. Currently, electricity usage is broadcast from an electricity meter to a portable in-home display, with the intent that when consumers can conveniently see their electricity usage, they will turn off any unnecessary appliances. In the foreseeable future, the display panel could morph into a control panel to control lights and other appliances. To ensure interoperability, all ZigBee devices need to run a complete software stack."

## Powerline Communication

"Another wired communication technology, called powerline communication, requires no dedicated wire. Command and control functions are sent and received on 110 or 220 V AC wires. Although many variations of powerline communications exist, frequency shift keying (FSK) is a popular choice because of the low data rate for command and control functions. In this method, high frequency is used for communication on a 60 Hz power line. The quality of powerline communication inevitably varies, and the error correction, such as forward error correction, reduces the error rate and improves the bandwidth of the medium. The data rate can achieve 30 kbps or even higher. In comparison to a regular switch, powerline communication requires an MCU to run a network stack. However, the benefits are tremendous, because the software stack makes the network expandable, thereby connecting and controlling every smart appliance plugged to the powerline. Utility companies have started evaluating powerline communication as a complementary technology to a radio technology called ZigBee for the Smart Energy Home Area Network."

# CHAPTER 10

# *PRACTICAL DESIGN OF AN LED FLASHLIGHT*

IN PREVIOUS chapters, we have reviewed all of the important properties of LEDs, their circuitry, and their physical and mechanical environment. Now it is time to put all of that information together. In the next four chapters we are going to design four complete devices from scratch: a battery-powered LED flashlight; a USB-powered reading light; an automotive tail light; and a 120 VAC-powered LED light bulb. These four devices give us opportunities to use much of the information that has been developed. We also add in a lot of design-specific new information, such as how to select components, how to perform testing, and how to calculate MTTF. In fact, you can think of Chapters 10–13 as being case studies in how to actually design LED lights.

The strategy of design we will use is typical of engineering projects. First we will find out what marketing says customers want. We'll then interpret this and refine it into the realm of what is physically and cost-effectively possible. We'll lay out specifications accordingly, and then check with marketing that this is what they had in mind. We will begin the actual design by making a first pass at the number and type of LEDs required, and estimating the power they require. We'll next select a control IC that can work electrically and also fit mechanically, and draw a schematic. Then we'll select real components from a distributor's website, to make the design manufacturable. We'll make a model of the mechanical environment, and use this to check that LED temperatures are acceptable. As a last step, we'll make a PCB layout. Throughout, we'll be making choices based on cost. Our final step will be to make a final estimate of the total cost of the design.

In this chapter, we'll be designing an LED flashlight. We'll follow our standard procedure as just outlined. But we're also going to add in a section on how to practically go about selecting components. When you go to a distributor's website, there are hundreds of similar components, and we want to know which ones are the best choices for the application—the general idea is to select the cheapest component that will work.

*Practical Lighting Design with LEDs*, Second Edition. Ron Lenk and Carol Lenk.
© 2017 by The Institute of Electrical and Electronics Engineers, Inc. Published 2017 by John Wiley & Sons, Inc.

## INITIAL MARKETING INPUT

For our first project, we're designing an LED flashlight. What marketing tells us they want in the first meeting is "A camping flashlight. Not one of those big police flashlights, something you can carry around. Get maybe 20–30 h on a couple of batteries, how about 'C' size? We want to sell it for about US$10 retail, get it mass produced in China for something like US$2.00." They show us Table 10.1 to explain the cost structure they want.

The first thing to note in this table is that there are a large number of intermediaries between the manufacturer and the ultimate consumer. After we finish designing this device, we're not going to make it ourselves. We're going to pass it over to our manufacturing partner in China. Almost everything produced in high volume is made in China these days. That's because they can do it much more cheaply there than anywhere else.

After the unit is made in China, it's sold to our company. We then use a representative, or rep (as they are known). We don't have enough sales people to visit all of the potential customers, so we hire someone whose job is to visit properly qualified customers. These reps have usually been in the business a long time, and have close ties with their customers. They can easily get in to see the buyer at the retailer. If you or I went to the retailer, we'd be politely shown the door. So it's beneficial for us to use the rep firm.

By the way, reps are not the only way to get sales. There is also distribution, known colloquially as distis. These are companies like Digikey or Mouser, which sell directly to original equipment manufacturers (OEMs). Our flashlight is going to be sold through retailers to consumers, so this doesn't concern us for the moment. The rep firm sells it to retailers, places like Best Buy or Sears. They in turn sell it to the end customer. So in the end, this flashlight goes through a lot of different hands before you go to the store and buy one. And of course, each of those people needs to be paid.

The Chinese manufacturer is supposed to produce the flashlight for $2.15. (This is the price in US$, converted from the Chinese currency, Yuan, at the current exchange rate.) Then they're going to sell it to our company for $2.54. That means their profit on the $2.54 is $2.54 − $2.15 = $0.39, which is 15% margin. Margin means the percentage of the sales price that is profit. So 15% of the sales price of $2.54 is $0.39, which is $0.39/$2.54 = 15% margin for them. Remember that margin is multiplicative, not additive! If their cost is $2.15 and they need 15% margin, then they have to sell it at $2.15/(1 − 15%) = $2.54, *not* at $2.15 + 15% = $2.47.

**TABLE 10.1  Marketing Cost Structure for an LED Flashlight**

|  | Price ($) | Margin (%) |
|---|---|---|
| Consumer | 9.99 |  |
| Retailer | 6.00 | 40 |
| Rep firm | 3.90 | 35 |
| US | 2.54 | 35 |
| Manufacturer | 2.15 | 15 |

We have bought the flashlight from China for $2.54. Our margin is going to be 35%. That's not that great for a US firm, but then we're not the first ones on the market with this type of device! So we have to sell it to the reps at $2.54/(1 − 35%) = $3.90. Again, we check that 65% of $3.90 is $2.54, so that was right, we get 35% of the sales price as profit.

Next, the rep firm has to sell it to the retailer. The rep firm of course expects to be paid, too. Their margins can range anywhere from 15 to 35%. Since we haven't used them much before, they don't know much about our products, or about our company. And so their commission is on the high end, 35%. As they get more familiar with us (and sell more of our products), their margins will drop (or we'll get a new rep firm!) To get that 35%, they need to sell the flashlight to the retailer at $3.90/(1 − 35%) = $6.00.

Now finally, the retailer is going to mark up the flashlight to sell it to the consumer. A retail outlet may want anywhere from 30 to 50% margin. Forty percent margin (which we're using here) means that the retailer gets the unit at 60% of the price they're selling it for: Forty percent of the sales price is their profit. (That is, gross profit. They have to pay rent for the store, they pay their employees, etc., so often their net profit—*real* profit—is only 10%.) So they are retailing it for $6.00/ (1 − 40%) = $10.00. ($9.99 is market-speak, it *looks* cheaper than $10.00.) And now you see why the device that cost only $2 to make in China costs you $10 in the store!

A little initial discussion with marketing reveals that what they have in mind for this flashlight is one of the newer types with a single LED in it. We suggest that there may be cheaper options: Perhaps it would be cheaper to have six or seven T1¾ LEDs in it? They ask us to cost this out also.

## INITIAL ANALYSIS

The first thing to do here is to scope out the battery, and see approximately how much power and then light output you can get from "a couple of 'C size batteries.[1]" We'll assume that "a couple" means two, because that is a common number of cells in a flashlight battery. (Of course, they are used in series.) A quick web search[2] suggests that a "C" cell has something like an ampacity[3] of 6 Ah at an average voltage of 1.2 V, which is 6 Ah × 1.2 V = 7.2 Wh. Since there are two cells, we will have about

---

[1] Each of the cylindrical items you buy from the store is properly referred to as a "cell." A battery is multiple cells connected together. But common parlance is to call individual cells "batteries" also, and we won't fight it.
[2] http://www.rayovac.com/~/media/Rayovac/Files/Product%20Guides/pg_battery.ashx Property of Spectrum Brands, Inc.
[3] Ampacity is a weird and not-very-accurate measure of how much energy a battery stores. It measures how many amps can be drawn from the battery for how many hours, and so its units are amp-hours (A-h.), or mA-h for the small batteries we're talking about here. It's hard to use ampacity for real calculations, because the battery voltage drops as it gets discharged. Worse, the battery voltage also drops in proportion to how much current is drawn, like a (nonlinear) resistor. When you connect a battery to a SMPS drawing constant power, more amps are drawn at lower voltage to keep the power constant. And thus as the battery gets close to being empty, the current rises rapidly, and the battery very quickly goes from "almost-empty" to empty. But ampacity is good enough for getting a rough guess at power. After all, there are only a few common sizes of batteries available.

14.4 Wh. At 25 h of operation (the average of marketing's specs), we have power available of 14.4 Wh/25 h = 576 mW. This presumably has to go through a converter. Assuming we can afford to achieve 85% driver efficiency on the $2.00 budget, the power available to the LED is 576 mW × 85% = 490 mW. We should be able to get a not-too-expensive LED mid-power device in the 100 Lm/W range. We'll lose maybe 10% in optics. Adding it all up we have 490 mW × 100 Lm/W × 90% = 44 lumens net. A little web searching suggests that 30 lumens is about right for a camping flashlight, so this is a bit high. We could instead extend the runtime to something like 25 h × (44 Lm/30 Lm) = 37 h, something to look at when we review the project with marketing.

Since this runtime is more than required, could we possibly get by with smaller (= cheaper) cells? "AA" cells have an ampacity of about 2.5 Ah, which at 1.2 V is 3.0 Wh. Two cells will be 6 Wh. So with this smaller battery at 25 h of operation, we have power available of 240 mW. Again assuming 85% efficiency, the power available to the LED is 204 mW. With a power device in the 100 Lm/W range, and 10% loss in optics, we will generate 18 lumens. This is on the low side, but we could reduce the runtime to 25 h × (18 Lm/30 Lm) = 15 h to get the 30 Lm, another option for marketing to consider.

We also told marketing about using multiple T1¾ devices. We visit www .mouser.com. Looking at the first five devices that list both brightness and viewing angle, we construct Table 10.2. Thirty lumens appear to be completely impractical with T1¾ devices because of cost. Even if the costs were OK, you couldn't fit 20 or 100 devices into a flashlight, even with surface mount.

Of course, these are just preliminary engineering estimates. We haven't accounted for temperature rise of the LED, which will reduce the light. But this is a second-order detail. What marketing really cares about is cost. We take a quick glance at Digikey, and find that the least expensive LED costs only a few cents. (We'll select the exact component in a later section in this chapter.) When we start putting together the other bits of the cost, we take a guess that the converter will be $0.50 and the housing will be $1.00. The difference between converting 500 mW and converting 1 W probably will make no difference to the cost of the converter. And heat-sinking an extra half watt isn't going to matter to the aluminum housing. So this is probably a safe estimate.

Let's put together a summary, Table 10.3, for marketing. We highlight the main choices, the brightness versus runtime, for both the "C" cells and the "AA" cells. To

**TABLE 10.2   A Few of the Possible 5 mm LEDs for the LED Flashlight**

| Device | Cost ea. (1 K pcs) | Intensity (mcd) | Viewing angle (°) | Lumens | Cost, 30 lumens |
|---|---|---|---|---|---|
| LTW-2S3D7 | $0.27 | 13000 | 15 | 0.7 | $11.57 |
| LTW-2E3C4 | $0.78 | 4200 | 15 | 0.2 | $117.00 |
| LTW-2H7C5S | $0.84 | 1800 | 56 | 1.3 | $19.38 |
| LTW-2V3C5 | $0.90 | 2600 | 30 | 0.6 | $45.00 |
| WP7114QWC/D | $1.09 | 3200 | 20 | 0.3 | $109.00 |

**TABLE 10.3   Battery Choices for the LED Flashlight**

| Option | 2 "C" cells | | 2 "AA"s cells | |
|---|---|---|---|---|
| Ampacity (Ah) | 6 | 6 | 2.5 | 2.5 |
| Mean voltage (V) | 2.4 | 2.4 | 2.4 | 2.4 |
| Energy (Wh) | 14.4 | 14.4 | 6.0 | 6.0 |
| Converter efficiency | 85% | 85% | 85% | 85% |
| Optical efficiency | 90% | 90% | 90% | 90% |
| LED energy (Wh) | 11.0 | 11.0 | 4.6 | 4.6 |
| LED power (W) | 0.44 | 0.30 | 0.18 | 0.31 |
| Hours (h) | 25 | 37 | 25 | 15 |
| Brightness (Lm) | 44 | 30 | 18 | 30 |
| Converter cost (US$) | $0.50 | $0.50 | $0.50 | $0.50 |
| Housing cost (US$) | $1.00 | $1.00 | $1.00 | $1.00 |
| LED cost (US$) | $0.10 | $0.10 | $0.10 | $0.10 |
| Total cost (US$) | $1.60 | $1.60 | $1.60 | $1.60 |

first order, there is no difference in cost for the different options. The choices come down to cell size and runtime.

The costs we haven't included here are things such as freight costs to ship the units from China, and import tariffs. And the costs listed here are pretty aggressive. But even so it seems likely that we will be able to meet the $2.15 price target. Of course, it will be up to us to make the $0.50 converter happen, but if it runs over there's some room to wiggle. Naturally, if it comes out to be less expensive to produce, our company can either sell it for less (increasing volume) or get better margin. Which one sales picks isn't of concern to engineering.

## SPECIFICATION

Marketing indicates that our overall choices are acceptable. In particular, operations says they can produce this device for $0.30 labor if the circuit board isn't too complicated. (Oops—we left out the labor cost and the PCB cost too!) Marketing decides to go with the "C" cell option since that doesn't affect the cost of the flashlight (batteries are not included). They also decide on longer not brighter, since customers understand hours better than lumens. We thus draw up a specification, Table 10.4, for marketing sign-off before the design begins.

There are a number of issues that are addressed in this specification that we have not yet explicitly considered. The first thing to understand is the meaning of the terms in the header, Min. (minimum), Typ. (typical), and Max. (maximum). In our conventions, "typical" numbers refer to design goals. It does NOT mean that the average unit will perform this way. And it does not imply that units will be tested to this spec. It just means that this is about what we expect. Thus, if there is only a typical number in the specification, such as for MTTF, then the parameter is a design goal, and is guaranteed by design not by production test. Conversely, any minimum or maximum specification is a guaranteed parameter, to be tested at production time

**TABLE 10.4   Specifications for LED Flashlight**

| Parameter | Min. | Typ. | Max. | Units |
|---|---|---|---|---|
| Brightness ($V_{bat} = 2.4$ V) | 30 | | | Lumens |
| CCT | | 5,000 | | K |
| CRI | | 70 | | |
| Battery size | | 2 'C' cell alkaline | | |
| Battery voltage (by design) | 1.8 | 2.4 | 3.1 | V |
| Run time | | 37 | | h |
| Ambient temperature (by design) | 10 | 25 | 40 | °C |
| Touch temperature | | | 60 | °C |
| Reverse battery protection | | No | | |
| MTTF | | 50,000 | | h |
| Shelf life | | 12 | | Months |
| BOM cost | | | $0.60 | US$ |
| Total cost | | | $2.15 | US$ |

unless otherwise noted. It is thus common in this sort of specification for there to be a lot of typical parameters, and few minima and maxima, as operations has to sign off that they can cost-effectively test these latter.

To start digging into this specification, we begin with brightness. A typical LED may have a minimum guaranteed flux at, say, 100 mA. While the minimum flux is guaranteed, the typical and maximum fluxes are sometimes not. So we're designing to the minimum, 30 Lm, not the typical. Of course, no one minds if it is brighter, so no maximum is specified. (In production there will be a maximum to be tested as well, to ensure that the converter is built correctly.) The other factor potentially affecting the brightness is the converter. We expect its efficiency to go down as battery voltage drops, since the current is higher. To avoid messy tests in production, brightness is specified at the typical battery voltage. That way, production can test the unit using just a 2.4 V power supply, rather than a battery, which would give varying results depending on its state of charge.

We've specified a CCT of 5000 K. Now, marketing didn't specify this parameter. We just felt that we had to specify *some* value for it. Otherwise, the manufacturer might use different LEDs in different weeks, as prices change. And then the light from the flashlight would look different, depending on when you bought it! But why this particular value? For one thing, most flashlights already have a high color temperature, so we should too. It's better for discriminating fine features in low light environments, such as reading. Further, higher CCT means higher efficacy, so runtime of the flashlight will be maximized by choosing high CCT. And the CRI? We don't really care. At such a high CCT, whether the CRI is 70 or 80—common numbers—won't be visible. Since we don't care, we just leave it as a typical number. As a general engineering principle, don't overspecify, only list the parameters that actually matter.

Maximum battery voltage is set by the design, having a two-cell alkaline battery. Note that this is an internal specification, not something that is tested on the production line. To show this, the specification states that this is a design specification. Minimum battery voltage is not determined by the battery, it could theoretically be as close to zero as you like. But practically, below about 0.9 V/cell there isn't much energy left. And the more the converter design becomes harder and expensive, the lower the voltage at which it must operate. We consequently decide to not have the converter work below 1.8 V. Of course, it should not blow up below that voltage, but operation is not guaranteed.

Runtime is of course by design, there's no way to test that in production. What we've done here is to use the 2 "C" cell spec, 14.4 Wh. Since our spec is only 30 lumens, with a 100 Lm/W part, we need only 300 mW into the LED. Dividing this power by the converter 85% efficiency, that's a power drain of 353 mW, corresponding to a runtime of 41 h. Finally, 10% optical loss gives us the 37 h runtime in the spec.

We first specified ambient temperature at a minimum of −40 °C, thinking that LEDs become more efficient at lower temperatures, and that it might get that cold when hiking in the winter. But a little reflection shows that there might be moisture and condensation issues on the PCB when coming indoors when the outdoor is below freezing. Further, electrolytic caps (if we end up with any) lose most of their capacitance below freezing point. So we end up specifying minimum ambient as 10 °C. Marketing reluctantly agrees to this based on cost, but wants us to make our next flashlight without this limitation. We agree to look at whether we can extend the ambient temperature range.

By "touch temperature" we mean the maximum temperature on the flashlight that can be accessed by the user. One would not like to get burned from a flashlight that's been on for 10 min! At 40 °C ambient, this means only a 20 °C rise. But the LED itself will be covered by a lens, and so the aluminum-to-air junction will meet this spec easily.

Reverse battery protection seems reasonable; lots of people put in batteries backward. However, the only small and inexpensive way to implement this is with a series diode, which is going to drop at least half a volt. This would take minimum voltage the converter sees down to 1.3 V, and thus raise the cost. Therefore, marketing decides to live without this feature.

MTTF has a design goal of 50,000 h. Of course, there is no way to test this on a production line, but a converter with only a few parts, plus an LED running not very hot make this seem achievable. In the end we will meet this specification (while we don't show the calculations here, a representative sample of the calculation is shown in the next chapter).

Finally, shelf life is something that marketing wants to put on the label. When the flashlight is off, they don't want the battery to be drained like in a car or cell phone. But all electronic circuits have *some* leakage, including of course the battery itself, which has self-leakage. What we end up with is that the flashlight will have a mechanical switch, so that it won't draw any current at all. Shelf life is then set by the shelf life of the battery.

## POWER CONVERSION

Having estimated the power needed for the LED, the next task is to design its power supply. We've made an initial allocation of power to the LED of 300 mW. The first thing we have to do is to translate this into current. Now white LEDs have a typical forward voltage of about 3 V at 25 °C, and this drops by about −3 mV/°C. For our initial design, we'll assume the junction of the LED in steady-state is going to be about 60 °C. We thus expect a forward voltage of about 3 V − [3 mV/°C × (60 − 25 °C)] = 2.9 V. Dividing this into the power, we get an LED design current of 103 mA. We'll adjust this when we have a better estimate of the actual LED temperature, forward voltage, and of course efficacy.

The first and most critical item to select for the driver is the IC. It will be the most expensive component of the driver, and thus determines our ability to achieve our $0.50 goal. To select the correctly priced IC, we introduce the authors' *Divide by Three* rule. It has turned out, over a very wide range of designs, that the price of components in mass production in China can be reliably estimated. To do this, you take the 1000 piece price at a distributor and divide it by 3. For example, the 1N4004, a common diode, is priced at a distributor at 4.3 ¢ for 1000 pieces. When taken to mass production in China, it is a good estimate that the price will be 4.3 ¢/3 = 1.4 ¢ (rounded off). As another example, a 100 µF, 100 V aluminum electrolytic capacitor is priced at 12.2 ¢ at the distributor at 1000 pieces; in mass production this will cost 4.1 ¢.

Why does this rule work so well? We can only make some guesses. The authors have observed that pretty much all component companies seem to have similar price versus volume curves. Further, there are a couple of different companies taking profits before you buy the part from the distributor (the same as the markups on your product). Perhaps these two things act together to smooth out individual differences in margins. In any case, the Divide by Three rule gives remarkably good results, often within 10% of the real mass production pricing. Of course, there are limitations, primarily for one-of-a-kind parts. We don't currently use it for LED pricing because each LED is a unique part sourced by only one company. But it won't be long before LEDs are such a commodity that the rule will probably apply to them as well. To design a power supply that will cost $0.50 in production, we will target a distributor component price of $1.50 at 1000 pieces.

Let's turn back to the IC, then. We look for ICs designed for driving a constant current from a DC supply. We limit ourselves to parts that will work down to at least 1.8 V. We also eliminate parts that work substantially below 1.8 V, since we don't want the converter to still operate and pull huge currents at, say, 1.1 V. (Of course, we could just design in a current limit, but then we've added extra cost for no good reason.) We eliminate some parts that are designed for voltage output. (You can tell these right away because their feedback voltage is measured in volts instead of millivolts.) We also eliminate parts that are obviously designed for nonpower LEDs.

Finally, we select the FAN5333A, a boost converter available in a nice small SOT23 package. It is priced at $0.70 at 1000 pieces, suggesting we can get it (or a similar Chinese part) in volume at $0.23, although in the actual BOM we will leave it at the 1000 piece price just in case. It works down to 1.8 V, has a feedback voltage of 110 mV, and has a 1 A switch built into it, conveniently eliminating the need for, and the cost of, an external transistor. Equally important for our design, it switches at 1.5 MHz, so that the inductor value, and thus size, will be small.

Our preliminary schematic is shown in Figure 10.1, taken directly from the datasheet. You might notice that we redrew the schematic to show the pins in the

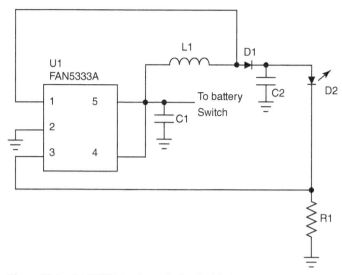

Figure 10.1    FAN5333A schematic for flashlight.

actual order they are physically in. That is, looking at the device on a board you would see pin 1 in the upper left-hand corner, and below that would be pin 2, and so on. The reason we like to draw schematics this way is that the connection lines provide a helpful guide for the PCB layout: Unavoidable trace crossings are flagged, which will be important if we end up with a single layer board. It's also convenient for the engineer when debugging a prototype.

We need to select initial values for the components L1, C1 and C2, D1, and R1. We'll adjust them as we progress further into the design. To start with R1, the feedback voltage to the FAN5333A is 110 mV nominally. To get 103 mA, we need R1 to be 110 mV/103 mA = 1.07 Ω. Power dissipation in this resistor will be (less than) 110 mV × 103 mA = 11 mW, and this is so small that any size resistor will work. The available resistance is of 1.07 Ω value, readily available in a 1% tolerance. (We use only 1% resistors these days, never 5%. There is no difference in cost.) It is available in sizes from 0402 on up. Since the current is an estimate anyway, we can go ahead and select this value for the time being. We can tweak the value during the more detailed analysis we make after LED selection.

The purpose of C1 is to buffer the battery impedance. While batteries don't change voltage very much when you pull moderate amounts of DC current from them, they have fairly high impedance at high frequencies. To see what this means, consider that the switching frequency of the FAN5333A is 1.5 MHz. Current is being drawn from the source by the inductor with a change in slope of the current twice every 1/1.5 MHz = 667 ns. If the impedance of the source for this current (namely, the battery) has high impedance at this frequency, the voltage that the inductor sees (at its input) will be less than the DC battery voltage—possibly a lot less. In fact, it will be $V_{battery} - (I_{inductor} \times Z_{battery})$. This will affect the operation of the converter. Even more importantly, that is also the voltage that is used to power the IC. If that voltage drops too low, the IC won't work anymore.

So the purpose of C1 is to provide high-frequency current to the converter. If the converter didn't have C1, it probably wouldn't work. To estimate how much capacitance we need for C1, we estimate the charge pulled from it during switching. The switching frequency is fixed at 1.5 MHz, so the on-time is less than 700 ns. During that time, we're pulling something like 100 mA. So the charge from the capacitor is less than $Q = I \times t = 700$ nec $\times 100$ mA $= 70$ nC. To get no more than 100 mV drop (say), we need $C = Q/V = 70$ nC$/100$ mV $= 700$ nF.

Now, the natural thing to do is to select a standard 1 μF part and move on. However, there's a problem. It turns out that capacitors aren't the ideal components shown in textbooks. Since they are physical devices with actual length and width and internal wiring, they actually have some inductance. So at some high-enough frequency, a capacitor stops being capacitive and starts being inductive! That defeats the purpose of having it there. A rule-of-thumb for the frequency at which this happens is $f = 1$ MHz/C (μF). Thus, a 1 μF capacitor works up to about 1 MHz, while a 10 μF capacitor only works up to about 100 kHz. But our switching frequency is 1.5 MHz! So, instead of a single 1 μF capacitor, we should select *two* capacitors, in parallel. We choose them to be 470 nF each (so that they're capacitive up to about 2 MHz). These are available at 6.3 V in packages as small as 0201. This sort of part we should expect to be less than $0.02 in volume, and so two of them won't break the bank.

The purpose of C2 is to buffer the switching current from the LEDs, so it sees approximately DC voltage. The calculation, to the level of accuracy needed, is the same as for C1. We pick the same capacitors as we used for C1 in order to minimize the number of different parts needed (and to minimize stuffing errors on the part of the manufacturer!)

D1 is the rectifier. Since we have 100 mA output at some duty cycle, we can pick a 1 A device. Making the diode's rated current smaller wouldn't help the price much, although we'll look more carefully when we get around to actual component selection. However, at 1.5 MHz, recovery time of the diode would play a major role in efficiency, as would any forward voltage. We therefore pick a Schottky diode to minimize both kinds of losses. We will try to choose a SOD123 package in order to be small, and since pretty much any reverse voltage will do (schottkies start at about 20 V, and even with 50% derating that's good enough), we will pick the device with the lowest forward voltage. Volume pricing for this sort of component should be about $0.04.

The final item to select will be L1. We've been a bit cavalier in selecting the capacitors and diode, since there isn't much of a cost penalty to pay if we oversize them. But inductors become large and expensive quickly with increasing size, and so we better estimate the current through L1. The duty cycle for a boost converter is $DC = 1 - (V_{in}/V_{out})$. The maximum duty cycle occurs for minimum input voltage, and the output voltage has to include the forward voltage of the rectifier, which for a Schottky carrying much less than its rated current might be 300 mV. Putting all these pieces together, we find $DC_{max} = 1 - [1.8 \text{ V}/(2.9 \text{ V} + 0.3 \text{ V})] = 44\%$. Peak current through the inductor is then the DC current plus half the peak-to-peak ripple current. Remember that the ripple current is how much the current changes from its average during a switching cycle. The factor of one-half is because the current in the inductor goes both above and below the average, and we want just the amount above to find the peak. The ripple current is given by the voltage across the inductor times how long that voltage is applied, divided by the inductance, because $V = L \, dI/dt$. We thus have $I_{pk} = I_{DC} + [V_{in} \times t_{on}/(2 \times L)] = 100 \, \text{mA} + 1.8 \, \text{V} \times [(44\%/1.5 \, \text{MHz})/(2 \times 10 \, \mu\text{H})] = 125 \, \text{mA}$. We took the $10 \, \mu\text{H}$ value from the recommendation of the datasheet.

## SELECTING COMPONENTS

So now we know approximately the component values we're going to need for this design. But we're not done yet, of course! In the first place, we need to select exactly which LED we're going to use, as this affects everything else in the design. And further, to actually *build* this design, it's not enough to know that R1 is supposed to be about $1.07 \, \Omega$. We need to select an actual component for manufacturing to procure and build with. Why not just let manufacturing select one? For one thing, there are other people involved with the design before it even gets to manufacturing. In particular, layout has to be able to design a PCB with the components. And what sizes are available for purchase for R1? Just because it's a normal 1% value doesn't necessarily mean it's stocked. What we need to construct is a BOM (Bill of Materials) that lists *exactly* which part numbers are to be used for the design—even resistors.

And what about the cost? Maybe we've picked an unusual value, and it costs $0.10 rather than a fraction of a cent. In that case, we will have to decide whether it is critical to have exactly that value, or whether we can go to a cheaper value without affecting performance unreasonably. That's not something manufacturing can (or should) decide. Although it's not a consideration for the components on this board, suppose that R1 had to be a 2 W resistor. Those can be expensive, and there are a limited number of resistance values available in that power rating. In that case, we might want to split it into two 1 W resistors. We would need to decide—again based on cost and availability—whether to put them in series or in parallel.

Even after we finish with the BOM, there may still be changes. In particular, layout may decide that there isn't room for all the big 0805 resistors that we might select, and may request us to reduce them to 0603 or 0402. Then we have to go back and recheck power ratings, tolerances, and so on. Layout requests change particularly frequently when doing a Metal Core PCB (MCPCB), since this type of PCB is usually a single-layer board. Any trace that is topologically a cross with another trace needs to go underneath a component. With some careful thought, making certain components bigger allows the trace to go under an already-existing component. If not, then a $0 \, \Omega$ jumper (which is a normal resistor with a value of $0 \, \Omega$) has to be added to the schematic and BOM to let the trace cross over. No, you can't just use a piece of wire, because how would that wire get attached to the board?

Let's start then with the key component—the LED. The only specifications so far are that it should be 30 lumens with a 100 Lm/W minimum efficacy, and should be about 5000 K CCT. We routinely use Digikey for selecting components. Of course, there are quite a number of other distis. But they all have approximately the same selection methods: You progressively narrow down what you want with dropdown menus or selection boxes. So going to Digikey's home page, we type in "LED."

Initially, there are a huge selection of products, most of which have nothing to do with what we're looking for, for example "sirens" and "rechargeable batteries." But we select "LED Lighting—White." This page turns out to have 19,508 choices!

Now, the first thing to do is to limit our selection to components that Digikey actually stocks. All distis stock a huge variety of common components. If they don't stock it, chances are reasonably good that it's not very popular. In that case, we probably don't want to use it, because how will we get it for production? So we check the box "In stock," which reduces the number of choices by a factor of 6, down to 2960. Now we know we wanted 5000 K. We go to the box listing available CCTs. There are a number of listings for 5000 K—those that are explicitly 5000 K, and those that give a range of CCTs with 5000 K being nominal. We'll choose all of them, except the last two that say "5000 K ~ 10,000 K" and "5000 K ~ 6500 K." These seem too wide. This reduces our choices by another factor of 7, leaving "only" 421 choices out of the original 19,508.

Anything else? Scrolling across all the selection boxes to see what else is there, we see "voltage–forward." There turn out to be LEDs with voltages as high as 48 V. So we recall that we assumed the forward voltage would be about 3 V. If the forward voltage was much higher than that, we'd need quite different components. So we choose only those LEDs with forward voltage below 3.4 V, leaving 288 devices. Which one to choose? Of course, the cheapest one that meets all the specs! Selecting the "up" arrow on "Unit Price" to sort by price with lowest price first, we find that the cheapest LED is from Luminus, part number MP-2016-1100-50-70. It's 5000 K, produces 27 lumens at a current of 60 mA (150 Lm/W!) and costs only 4.2 ¢. How times have changed! In the first edition of this book, 5 years earlier, the best LED we found was 100 Lm/W and cost $2.98! Of course, using this LED now takes a lot less current than we had planned on; let's quickly adjust the component values.

R1 was 1.07 Ω; since the current is reduced to 60 mA, we will change it to 110 mV/60 mA = 1.83 Ω. And since it's less current, it's also less power, so we don't need to recalculate that. For the input and output capacitors, our previous calculation is changed to $Q = I \times t = 700$ nec $\times 60$ mA $= 42$ nC, and $C = Q/V = 42$ nC/100 mV = 420 nF. So now we could probably get by with a single 470 nF for each, but we'll leave in two to improve the performance. If we run into space or cost issues at the end, we can remove the extra caps. The diode sees less current, but that doesn't affect us since we haven't picked one yet. The inductor value is 10 µH no matter what, but the peak current is reduced to $I_{pk} = 60$ mA $+ 1.8$ V $\times$ [(44%/1.5 MHz)/ $(2 \times 10$ µH)] $= 85$ mA.

Having selected the LED, next we are going to look at all the non-LED components. We start with R1, just now recalculated to be 1.83 Ω. The same ideas that we're about to use here hold equally well for all the other components. Now, maybe it seems easy: pick a resistor, how hard is that? But there are actually a huge selection of choices available. So here's our strategy. We're going to select a resistor of (approximately—remember our comments above about not overconstraining the design) the right value. Then we're going to add additional constraints, one at a time, to narrow down the choices. The constraints are going to be things we have to have—not just random choices. And then, when we have our final list of choices that will meet our minimum requirements, we're going to select the cheapest one. This then gives us the cheapest component that will work in the design.

So to get started, on Digikey's main web page we type in "resistor" as the keyword. This gives a long list of things that (presumably) have the word resistor in their title, such as "cable assemblies" and "filters." Under the heading "resistors" we see there are multiple types, such as "chassis mount" and "chip resistors"; of course we pick the latter. When we come on this page, we immediately notice that there are 364,577 different surface mount resistors available (we're not making this up!). So chances are that anything we want is available. The hard part is rather narrowing it down.

Just as before, the first thing to do is to put a check in the box "In stock." Unless it's something that we're willing to go to the manufacturer for, and probably wait 12 weeks for them to make, we only want parts that are readily available. Applying this filter cuts the number of possible selections by 2/3rd, down to 128,408. There's still a way to go. We recall our discussion above that the power dissipated in R1 is minimal. So, the only important parameter is the resistance value, not the power rating. However, there isn't a $1.83\,\Omega$ available. The closest value shown in the list of resistance values is $1.82\,\Omega$, which is within 1% of the calculated value. But to ensure we don't end up with a value that is unnecessarily expensive, we select a value on the other side of our desired value as well, $1.87\,\Omega$. That's a couple percent high, which would probably be acceptable for this design.

We're now down to 43 choices. One more thing that we wanted was to use only 1% resistors. While this is a good general policy, it also makes sense to do it for this resistor in particular. If the resistor tolerance was 5%, the current through the LEDs could also be this far off, and that would affect everything in the design. As it turns out, the two values we picked, $1.83\,\Omega$ and $1.87\,\Omega$, are only available in 1%.

Now at this point, we could, for example, specify the size of the device. It's available from 0402 up through 1206. But unless the board is very tight on space, we shouldn't overspecify. Similarly, power ratings are available from 1/16 to 1/4 W. But since power dissipation is very low, we don't need to specify that either. Instead, let's just see what is cheapest.

By clicking on the "up" arrow under the column "Unit Price," we get a sorted list. The first, and the cheapest, resistor is $1.82\,\Omega$ in a 0402 package. Its cost is $0.0054, less than 1 ¢, which is negligible. So we select this part, carefully noting down the Digikey part number, 541-1.82LLTR-ND, as we will want to include that in our BOM.

So that took a fair amount of doing just to select a resistor. But, of course, selecting the exactly right component is an important part of engineering. And the reality is that after you've done it for a while, the selection becomes routine and goes very quickly. Of course, each component type has its own selection criteria. You just need to get some practice at it and it comes naturally.

The next component we need to choose is a capacitor, the input and output capacitors, which we decided to make the same. Recall that we decided that each should be two 470 nF caps. So we go back to the Digikey front page, and type in "capacitor." After selecting "ceramic capacitor" from a dozen different types of capacitor, and selecting "in stock," we end up with 74,446 choices. We don't bother with values that are close—experience has taught us that 470 is a standard value. Indeed, that's why we picked it. So selecting 0.47 μF (they don't use nanofarads on their website), we have 1920 choices.

Now, for capacitors, differently than for resistors, the tolerance usually doesn't matter that much. The capacitor is just for bypassing high-frequency switching noise. The normal ceramic capacitor has 10 or 20% tolerance, and for a filter capacitor it doesn't much matter. In fact, the only thing that much matters is the voltage rating. The capacitors are available from 4 V to 1 kV. Now, the battery can be as high as 3.1 V. We could select a 4 V capacitor (which will probably be the smallest), but a little extra safety margin wouldn't hurt. So the lowest voltage rating we look at will be 6.3 V. But even though that's plenty, we know that capacitors less than 1 μF often have voltage ratings as high as 35 V or even 50 V, because it doesn't cost the manufacturer anything extra to get the higher voltage rating. So to avoid over-specification, we select the voltage rating to be anything from 6.3 to 50 V. We also add in the selection that the capacitor should be surface-mount.

We still have 1376 choices. Let's just pick the cheapest one. Sorting again by cost, the cheapest turns out to be a 50 V part in 0603. However, reviewing the list of parameters for the part, the tolerance just happens to turn out to be −20%, +80%. But +80% would take us up close to 1 μF, and we said that that was too large to work properly. So we go back to the selection table and choose only the 5, 10, and 20% parts. This leaves 1311 choices (only a few parts got cut out, because for ceramic capacitors the −20%, +80% tolerance is unusual). And resorting by cost, the cheapest is a 6.3 V in 0402, Digikey part number 1276-1479-2-ND, which we select.

The third component to select is the diode D1. We've already decided that it ought to be a Schottky, both because of the high switching frequency, and also to minimize power loss. By now the routine is familiar. We go to the front page of Digikey, and type in "Schottky." After selecting "Diodes, Rectifiers – Single," and "in stock" and "surface mount," we have 4955 choices left. Since we said we have 60 mA of output, our diode should have a current rating of at least 100 mA. And since we know that diodes come up to at least 1 A without much cost or size penalty, we choose values between these two limits, leaving 2136 choices.

There's one more choice important for diodes, their reverse voltage rating. Now for non-Schottky diodes, we would want to choose a voltage rating at least 30–40% higher than the highest voltage the diode should ever see. But Schottky diodes are special. They have very considerable leakage current when the voltage applied from cathode to anode is close to their rated voltage. So as a rule-of-thumb, the voltage rating for a Schottky should be *double* the maximum expected reverse voltage. As it turns out, in our case, the reverse voltage is only the 3 V the LED takes, so any Schottky will do. They're all more than 20 V. Sorting by price, the cheapest is a 30 V, 200 mA Schottky in a SOT-23. But looking at its forward voltage, we see that it has 800 mV at 100 mA—no better than a regular diode!

Going back up to the "forward voltage" selector, we see values all the way from 220 mV to 1 V. We want something with good efficiency, so let's choose a forward voltage of not more than, let's say, 350 mV. This leaves us with only 48 choices. The cheapest is a 30 V, 500 mA diode, rated at 340 mV at 100 mA for 6 ¢. If we picked a diode with higher current rating, it might have rather less voltage at 100 mA, so we scroll down a bit. We come to a 20 V, 1 A diode, rated at 340 mV at 1 A for 9 ¢. Opening the datasheet, we see that at 100 mA its *I–V* curve shows 275 mV max. Is saving 340 mV − 275 mV = 65 mV worth an extra 3 ¢? The power dissipation saved will be 65 mV × 100 mA = 6 mW. We probably don't care about this, and so we end up with the first diode, Digikey part number CTS05S30L3FTR-ND. The *I–V* curve in the datasheet says it has a forward voltage of about 250 mV at 60 mA.

The final component to select is the inductor. Recall that we followed the datasheet selection for the value of 10 µH, and that the current rating should be at least 85 mA. Typing in inductor on the front page and selecting "fixed inductors," we see 122,994 choices! But it turns out that most inductors are not stocked, so selecting "in stock" narrows our choices down to 46,451. Our first selection, then, is inductance. Although 10 µH is a standard value, there are lots of close-by values it would be silly to ignore. Being off a little probably won't affect things, so we look for inductors between 9 and 11 µH, and also choose surface mount, leaving 2385 choices.

Next we need to find the current rating. There are two kinds of current ratings for inductors. "Current rating" *per se* refers to how much RMS current can be put through the inductor without overheating it (it's usually the current that produces a 40 °C temperature rise). "Saturation current" is the maximum current that can instantaneously be put into the inductor without degrading the inductance. So we need to select both: We choose minima of 125 mA for the current rating and 150 mA for the saturation current rating, with the upper end of both 500 mA, leaving us 72 choices. The least expensive component, Digikey Part Number 445-6396-2-ND and manufacturer Part Number MLZ2012M100, is 10 µH at 350 mA current rating and saturation current of 200 mA.[4] It's nice and small (0805), and inexpensive ($0.052). Resistance is only 470 mΩ, so resistive power dissipation will be only 6 mW, and so we choose this one.

The last thing missing is the Digikey part number for the IC. So we type in "FAN5333" into the front page. Something unexpected appears: There is both an FAN5333A and an FAN5333B! And while the "A" version we were planning on is 63 ¢ at 3000 pieces, the "B" version is only 40.6 ¢! Scrolling across the web page, there doesn't seem to be any difference at all between the two. So what's different? When we open up the datasheet again, the only difference between "A" and "B" turns out to be the current sense voltage: The "A" version is 110 mV, as we assumed, while the "B" version is 315 mV. If we change, that will change the value of R1 (and possibly its power?) and the output voltage of the converter. But for 22 ¢? Of course we're going to do that! So we redo the calculation for R1: 315 mV/60 mA = 5.25 Ω, and power dissipation will be less than 60 mA × 315 mV = 19 mW. Looking up on Digikey, we settle on 5.23 Ω, 1% in a 0603 package, at a cost of $0.002.

---

[4] So you might wonder, how can the current that can be put into the inductor without overheating, the rated current, be less than the current at which the inductor saturates? There's actually a couple of things going on here. The inductor doesn't hard saturate, going instantaneously from rated inductance to zero when you exceed the saturation current. Saturation current is usually where the inductance drops 10 or 30% from its nominal value. Furthermore, the rated current is based on the power dissipation in the winding, which depends on the RMS current. If you have very high amplitude current spikes for a short time, the RMS can be very high even though the average is low. So this can overheat the inductor, even though it only momentarily saturates.

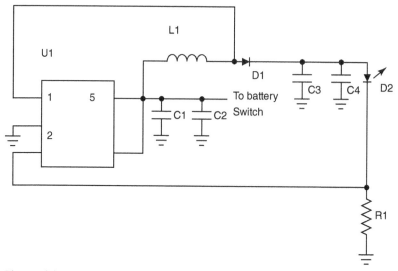

Figure 10.2 FAN5333A final schematic for flashlight.

Our schematic ends up as shown in Figure 10.2. Estimated cost is shown in Table 10.5. To be safe, the production cost of the LED and IC have been left at the same price as for 1000 pieces, since they are unique and thus hard to find cheaper Chinese substitutes for. Perhaps the price will be 10% lower when we get to volume, it's hard to guess. The only thing in the pricing table that we haven't mentioned yet is pricing on the PCB. We're going to assume that the whole thing fits on a circular PCB 1 in. in diameter. That's so it fits inside the cylindrical body of the flashlight. The area

**TABLE 10.5 Final BOM for LED Flashlight**

| Ref des | Description | Digikey part # | Mfr. part # | 1Kpc price | Est. -rice |
|---|---|---|---|---|---|
| C1 | 470 nF, 6.3 V, 0402 | 1276-1479-2-ND | | $0.014 | $0.005 |
| C2 | 470 nF, 6.3 V, 0402 | 1276-1479-2-ND | | $0.014 | $0.005 |
| C3 | 470 nF, 6.3 V, 0402 | 1276-1479-2-ND | | $0.014 | $0.005 |
| C4 | 470 nF, 6.3 V, 0402 | 1276-1479-2-ND | | $0.014 | $0.005 |
| D1 | 500 mA, 30 V Schottky | CTS05S30L3FTR-ND | CTS05S30 | $0.066 | $0.022 |
| D2 | XNOVA LED, 5000 K, 0806 | 1214-1209-1-ND | MP-2016-1100-50-70 | $0.042 | $0.042 |
| L1 | 10 μH, 360 mΩ | MLZ2012M100 | MLZ2012M100 | $0.065 | $0.022 |
| R1 | 5.23 Ω, 0603 | 311-5.23HRTR-ND | | $0.002 | $0.001 |
| U1 | Controller | FAN5333B | FAN5333B | $0.406 | $0.406 |
| PCB | 1 in. diameter | | | | $0.075 |
| Total | | | | | $0.588 |

of the PCB is $\pi \times (1 \text{ in.}/2)^2 = \frac{3}{4} \text{ in.}^2$. A good rule-of-thumb for volume PCB pricing is 10 ¢/in.$^2$, so that we have \$0.075 for the (volume) pricing for the PCB.

So, the cost for the power supply BOM ends up at \$0.60, including the PCB and the LED. Our original goal was \$0.50 for the power supply and \$0.10 for the LED, so we're smack on target. It's really quite remarkable that you can build a complete power supply for under 40 ¢! We still don't have a good handle on the other costs such as assembly cost, wiring cost, and the real cost of the mechanical part of the design, but we've done our (electrical) part of the job well. We also don't know how much cost there might be in incoming inspection from the manufacturer, and in freight charges. Of course, we'll have to see how the Rule of Three works out in actual production, but it's looking very promising. The most expensive part of the whole design is the IC. We're committing at this point—we can't switch to a different IC without a major redesign. Having a buyer concentrating on getting the best possible price for this will be the best way of further lowering the cost.

Let's do a couple of quick design checks before finishing up. The battery voltage can be as high as 3.10 V (=2 × 1.55 V/cell). According to the datasheet, the LED voltage at 60 mA will be about 3.0 V. The output voltage is *lower* than the input voltage! Are we violating the condition for a boost converter to work (output voltage must be higher than input voltage)? Actually we aren't. Remember that R1 has a voltage drop of 315 mV across it, and D1 also has 250 mV. Finally, L1 at 360 mΩ at 60 mA has another 20 mV drop. Adding it all up, the equivalent output voltage is (3.0 V + 0.32 V + 0.25 V + 0.02 V) = 3.59 V, which is substantially higher than the input. What about the duty cycle? The duty cycle is 1 − 3.10 V/3.59 V = 14%. So the transistor will be on for 667 ns × 14% = 91 ns, which is probably OK. Some devices have a "blanking time," a time during which the turn off of the MOSFET is inhibited because of noise. But the FAN5333 doesn't mention such a thing, so the short on-time is probably OK. And in any case, if the duty cycle is a bit longer than is needed, the LED will just be a little brighter for a while, until the battery voltage starts to decline. The battery impedance and the *I–V* curves of the diode and LED will ensure that nothing catastrophic happens, as the battery voltage will decline with increased power, while the LED forward voltage will increase.

## THERMAL MODEL

Since we've got a power dissipation estimate for the LED (60 mA × 3.0 V = 180 mW), we can estimate the LED temperature. To do this, we need to create a thermal model for the flashlight. What this means specifically is determining the thermal resistance of the path (or in some design paths) from the LED to the ambient.

The first link in the path from the LED to the ambient is from the LED junction to the solder point. The datasheet specified that (at 60 mA, conveniently) this is typically 35 K/W (°C/W and K/W are interchangeable). We need to know this number because the lifetime of the LED depends on its junction temperature, and not its package temperature.

The next link in the thermal path is the way the LED is mounted. For this design, we have opted to use a MCPCB. As recently as the first edition of this book, it had been customary to use FR4, because of the cost. But nowadays, the cost of MCPCB is essentially the same as that of FR4 in mass production. The thermal resistance of metal-core boards is significantly lower than that of FR4. For example, consider an FR4 PCB with plated-through holes. Suppose we laid out the board to have a dozen such holes. Following the lead shown in the Lumileds' Application Brief,[5] we estimate the thermal resistance to the backside of the PCB to be (# holes in Lumileds' design/# holes in our design) × Lumileds' thermal resistance = $(33/12) \times 7$ K/W = 19 K/W.

Now contrast that with a MCPCB's thermal resistance. Typical material is 1 W/(m K) (and 3 W/(m K) isn't much of a price premium). The board area we already estimated at ¾ in.$^2$ = 0.0005 m$^2$. The typical thickness of a MCPCB is 0.062 in. = 0.0016 m. Remember that the thermal conductance of a material is the thermal conductivity times the area, divided by the thickness of the material. So, in this case the thermal conductivity of the board is going to be 1 W/(m K) × 0.0005 m$^2$/0.0016 m = 0.31 W/K, giving a thermal resistance of 1/0.31 W/K = 3.2 K/W. That's a factor of 6 better than the FR4.

We've taken the LED thermal path to the backside of the mounting PCB. For the next step in the thermal path, the PCB is glued to a small piece of aluminum. When we talk to manufacturing, they intend to use 'some epoxy' for gluing. So we take the number for a typical epoxy in the same Lumileds' document, 0.8 W/(m K) for the thermal conductivity. The thickness of the glue we presume will be the 100 μm Lumileds uses. We have A = $\pi (1$ in.$)^2/4$ = 0.239 in.$^2$ = $1.5 \times 10^{-4}$ m$^2$. We thus end up with thermal resistance through the epoxy of $10^{-4}$ m/(0.8 W/(m K) × $1.5 \times 10^{-4}$ m$^2$) = 0.8 K/W.

Thermal conductivity of aluminum is very high, something like 100 W/(m K). Further, all the aluminum pieces in this design are relatively thick. So we're just going to approximate the thermal resistance of the aluminum as zero. So we've gotten from the thermal pad of the LED to the aluminum case of the flashlight. But wait! There's one more thermal resistance in the model. That's the resistance from the case to the air. As discussed in Chapter 6, the surrounding air is not really an infinite heat sink. It conducts heat through both radiation and convection. Which one dominates depends on the details of the arrangement, and takes a lot of effort to compute. For practical purposes, we will use the estimation method of that chapter. The power dissipation is 180 mW, and the surface area of the flashlight is [2 × π × (½ in.)$^2$] + (π × 1 in. × 8 in.) = 26.7 in.$^2$ (two circular ends, plus a cylinder that is 1 in. diameter and 8 in. long), so the power density on the surface is 0.007 W/in$^2$. Using a linear approximation, this is a 1.0 °C temperature rise, corresponding to a thermal resistance of 5.5 °C/W.

Our completed thermal model is shown in Figure 10.3. We can now compute the typical LED temperature for a lifetime calculation. Typical LED occurs for typical ambient temperature of 25 °C. We have LED temperature = ambient temperature + (power × total thermal resistance) = 25 °C + [180 mW × (35 + 3.2 + 0.8 + 0 + 5.5°K/W) = 33 °C. This small temperature rise means that we don't have to worry about LED lifetime, and thus about MTTF. Clearly, even if we ran the LED at 300 mW, it wouldn't be much hotter.

[5] Application Brief AB32, Lumileds, 10/08.

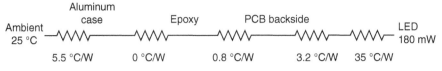

Figure 10.3 Thermal model.

The other temperature we need is the maximum temperature of the aluminum, because of the touch temperature requirement. This occurs when the ambient is maximal, 40 °C. For this, the PCB and epoxy part of the string of thermal resistances don't matter. We have aluminum temperature = ambient temperature + (power × air thermal resistance) = 40 °C + (180 mW × 5.5 K/W) = 41 °C. Again, we are well below spec.

Before we leave the thermal model, we should comment on two other thermal paths from the LED. Of course, for this flashlight, additional paths won't matter, since the temperature is already low. But in other designs they might matter, so a brief mention is in order. We already know from the Chapter 6 that the air surrounding the LED is not very important for thermals in many designs. Indeed, for normal LEDs, the thermal resistance for this path is not even specified. And in any case, there is a lens on top of the circuit board, so any air circulating will not have a direct path to ambient.

A potentially more interesting thermal path is through the wires that connect the PCB to the switch and battery. We can estimate the thermal resistance of this path the same way as all the others. Thermal resistance = one over the thermal conductivity of copper times the length of the wire divided by the cross-sectional area of the wire. Using 2 in. of AWG20 wire, we have $\Theta_R = 0.05$ m/(400 W/(m K) × 5 × $10^{-7}$ m) = 250 K/W. We have both a positive and a negative wire, which are thermally in parallel, giving 125 K/W, but clearly in this case this is insignificant compared to the main thermal path for the LEDs. (Remember that the thermal resistance for this path and the main one are in parallel, so a large thermal resistance in parallel with a small one is unimportant.)

# PCB

Since all the thermal and performance numbers look good in this design, we won't need to go back and revisit our assumptions. In particular, the typical temperature of the LED is lower than we assumed and so the light output will be slightly higher than we estimated. No one minds if it's a little brighter, only if it turned out to be dimmer. If we were intent on squeezing out the last bit of performance from the batteries, we could re-estimate the light output and scale back the current from the driver. But this is likely to be a few percent savings at most, and we forego the redesign in favor of getting the flashlight to market on time.

The last step in this process is to design the PCB. Since there is only one LED on the LED PCB, we will simply mount it on the same board as the driver, rather than having a driver board and a separate LED board. (This saves the cost of the extra board, as well as some cost in wiring.) So the very first step is to turn over the schematic to the layout engineer to capture in a layout program. Sometimes this is the same person as the design engineer. But it's much better if you can have an engineer whose specialty is designing boards, since there are so many special considerations in layout. It's almost as complicated as the electrical design! And there are even specialties within layout. Just like there are design engineers who specialize in LED lighting or power, there are also layout engineers who specialize in these areas as well.

We're frequently asked, "Can't you just have a person capture the schematic to a computer, and then push a button and the computer lays it all out?" Well, maybe that sort of works for digital design, where you have 32-bit wide buses, and all the traces run in parallel. Even then there seems to be plenty of work for the layout engineer. They have to enter all sorts of data about impedances, adjusting this and that. But with analog design, we're not aware of any program that can do a reasonable job. There are minimum spacings between traces for voltage (not on this design, but there will be on AC designs in a later chapter). You have to consider trace width for current carrying capacity, and also for voltage drop. You need copper around hot components to spread out the heat, and so on. So this really takes a person to work on it, and it can take quite a bit of time.

The schematic as captured in the computer program is shown in Figure 10.4 (we use DesignSpark, which is free on the web). It looks a lot like our hand-drawn schematic, although neater. It also shows all the reference designators, and the values and package sizes of all of the components. We feel it's easier to debug the board if there's a problem, if the schematic contains all this information. Note that the layout engineer has converted the reference designator of the LED, which we called D2, to a more standard LED1. You might also note how we have numbered all of the components. Numbers start at the top-left corner, and to the extent possible, increase sequentially toward the lower right. We do this again for debugging—if there's a problem with C4, it's easier to find if the numbering is done properly. If the schematic has to be revised, as it frequently does, we will renumber all the parts to maintain the sequence.

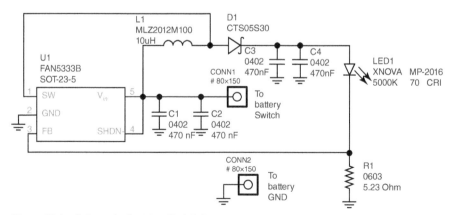

Figure 10.4   Schematic for LED flashlight.

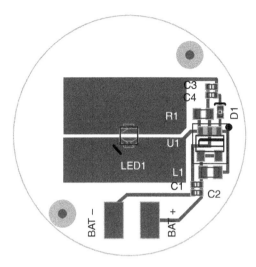

Figure 10.5   Layout of LED
flashlight. (See color plate section.)

One more thing to note on the schematic is the presence of the connectors. Although mechanical connectors are often used on electronic assemblies, in this case we are just going to solder wires to pads, because of cost. So these are just pads. But we still create a symbol (and footprint, see below) for them, as they have to be present in the actual layout. One of the connectors goes to battery ground, the other goes to the switch that turns the flashlight on and off.

Our layout is shown in Figure 10.5. The board is circular with a diameter of 1 in. (Yes, manufacturers will make round boards. It doesn't cost much extra.) It is obvious that with so few components in the design, the board could have been substantially smaller. The first thing to notice in this layout is that the colors aren't the real colors you get when you fabricate the board. The colors are selected to make it easy to see on the computer screen. For example, the red in the layout is the copper. You won't actually see the copper on the manufactured PCB, it's covered up because copper tarnishes. But we need to see it on the layout to ensure everything is hooked up correctly.

The copper is mostly covered by solder mask, which is a layer of electrically insulating polymer. The mask is usually green on most PC boards. However, for LED boards, the mask is white to help reflect light and not give a green hue to the light. This solder mask has gaps in it to reveal the copper for the metal pads to which the components are soldered. Again, on the actual PCB, these pads are not copper colored because they are tinned, they will appear silver. The pads are outlined in green in our layout, such as the BAT pad and the pins of U1. Note that on U1 the green outlines touch. That doesn't mean the IC pins are shorted, because the green is just a visual aid.

The component outlines are shown in blue, but will also not actually be on the physical PCB. The board outline is in green. Also observe that the reference designator is placed next to every component. This is the silkscreen layer and is black in our layout. We orient them all the same way for readability, but of course that's not always feasible. And if the board is packed with components, sometimes the reference designators may be not immediately adjacent to the component. One other choice would be to make the font smaller. In any case, the reference designators are for the engineers. The pick and place machines used to populate the PCB don't need them.

Turning now to the LED, note that it has a considerably larger solder pad footprint around it than the package itself might suggest. This is based on the datasheet's recommendation. One of the jobs of the layout engineer is to go through the datasheet of each (new) component and build a suitable footprint for it to be soldered to. These footprints are maintained in a library of footprints, so that you don't have to rebuild the same thing every time you use it. Companies that have multiple people doing layout may even have a controlled, central library for component footprints, to ensure that they are always the same. The reason is clear. If the footprint is wrong, when it's time to assemble the PCB, the components won't fit!

On our particular pad for this LED, you can see multiple green rectangles overlaid. This was necessary because the LED datasheet recommended an unusual shape for the pad, and DesignSpark only allows rectangles for pads. Later on, when we review the output Gerber files, we will see that the solder mask layer does indeed have the right pad shape for the LEDs.

Around the actual footprint of the LED extend (relatively) massive copper pours. These are to get the heat out of the LED, by providing a metallic conductor (copper) to spread the heat generated by the LED. Of course, in this case the LED isn't going to be dissipating much power, but it's good practice anyway. The other components don't dissipate any significant power, and so don't have any copper pours around them.

Since we were just talking about the connectors, we observe that they are just pads. The pads are relatively large, since wire has to be soldered to them. You have to assume that the soldering is done by hand, in which case there will be some tolerance in placement. They are thus separated by some distance, to ensure the manufacturer doesn't accidentally occasionally get them touching each other, which would short out the battery.

C1 and C2, the input bypass capacitors, are located right near the IC. These connections then go the shortest route possible to pins 4 and 5 of the IC for plus, and pin 2 for minus. We wanted the capacitor to be as close as possible to these pins, since this is where all the high-frequency action is taking place. The current sense resistor R1 is placed right next to the current sense pin of the IC and its ground, pins 2 and 3. This minimizes the chance of noise pickup into the current sense circuit of the IC. The return connection from the LED is also right there, again to minimize noise pickup. L1 and D1 are also placed very close to the IC, again to minimize the possibility of interference. And finally, C3 and C4, the LED smoothing caps, are next to the positive connection of the LED, and close to ground. The large copper pours around the LED provide a low impedance path for the current, and so the placement of these two can be anywhere close by.

You might notice that the LED has a diagonal black bar next to it. This is to show the cathode. This makes it easy for an engineer to check that the LED works—or even that it was installed correctly—by applying a DVM to the device without having to study the schematic. A five-pin IC such as we have in this design can't be put in backward, but an eight-pin easily could, and sometimes is! U1 thus has a dot in the silk screen to show where pin 1 is. D1 also has a bar, right next to its cathode. And if we had any electrolytic capacitors, we would also put a line mark next to the cathode (the negative), which matches the black line on the capacitor body that marks its cathode, in order to be able to quickly check its polarity. Adding all of these markings is standard practice.

So the basic idea of this layout was to make it as compact as possible, since the IC switching frequency is so high. The sense resistor in particular was placed right next to the IC pins, both signal and ground. The layout ended up wrapping the power components around somewhat more than the schematic showed in order to accomplish the tight packing. Maybe the schematic should have considered this, but we will leave it as is. This wrapping is of course a limitation of having a single-layer board. In fact, laying out a single-layer board is much more challenging than when a multilayer board can be used, because you have to move things around (sometimes a lot!) to find a way to make all the connections without needing a jumper. But in the case of LED boards, MCPCB is standard because of its thermal conductivity, and multilayer MCPCB is much more expensive than single-layer MCPCB.

One final thing to observe on this layout is the presence of the two "bull's eyes." These are "fiducials." They are tinned copper circles with no solder mask around them. The pick-and-place machine that places the components onto the board for soldering needs to know where the origin of the coordinates is. So it has a visual inspection tool that finds these fiducials, and uses them as reference points. Manufacturers prefer fiducials to be at opposite corners of a PCB, and two of them is preferred. Although they seem big relative to all the other markings on the board, we've almost never encountered a situation in which they absolutely can't be fitted on.

## DESIGN RULE CHECK

The final step before exporting the PCB design is to perform a Design Rule Check (DRC). Design Rules include specifications such as minimum spacing between traces, between components, between trace and board edge, minimum trace width, net completion, and a whole host of other rules. The most important aspect about DRC is to use it and ensure the design passes all rules before it's released. In our case, the DRC reports no errors, so we can proceed.

## GERBER FILES

Now it's time to export the layout into Gerber files that a PCB manufacturer can use. Gerber is a file standard and the layout design software will simply generate the files. DesignSpark uses descriptive language for the file names such as follows:

Flashlight – Top Copper.gbr

Flashlight – Top Solder Mask.gbr

Flashlight – Top Copper (Paste).gbr

Flashlight – Top Silkscreen.gbr

Flashlight – Drill Data [Through Hole] (Unplated).drl

Flashlight – Drill Ident Drawing [Through Hole].gbr

You will need Gerber Viewer software to actually import the files and view them. But you *should* view the gerbers because once in a while the layout software produces an error. In particular, sometimes a copper pour will leave a sliver of copper somewhere, even though the layout doesn't show it. This can accidentally short something out. And once, we saw a circle in the layout become an ellipse on the gerbers! The board manufacturer didn't know what to do. On a multilayer board, it's really hard to find such things.

The copper layer simply shows the areas where copper should remain after etching the rest of the copper away. The solder mask layer is a negative layer, meaning that the areas for pads where copper needs to be exposed are masked off such that the actual dielectric layer will not be applied there. It works similar to a masking tape for painting. The paste layer is a negative layer as well. This layer is made into a stencil with holes cut out where solder paste should be applied. During assembly, the stencil is placed over the PCB and paste is then applied and squeegeed off, leaving paste only in the appropriate areas. The unplated drill data file includes mounting holes and the board outline. Some PCB manufacturers want you to add the board outline to the top copper layer. We prefer to make an additional Gerber file named "Board Outline.gbr." If you have through-hole components or vias, then an additional plated drill file will need to be generated.

The Drill Identification Drawing Gerber file not only shows the location of the drill holes but also identifies their diameter via a drill tool size chart. A drawing file can also show various notes that are communicated to the PCB manufacturer, such as dimensions. In DesignSpark we create a drawing layer that is added to the Drill Identification Drawing Gerber file. In the next section on panelization, you will see an example of how it's used.

Again, it is highly recommended that you review your Gerber files using a Gerber viewer to verify that the Gerber export worked as expected and that your footprints are properly made. Gerber viewers are also useful for checking your board for manufacturability. They generally have tools for checking for minimum silkscreen width, minimum solder mask width, and so on. These specifications should match the capabilities of your PCB manufacturer.

## PANELIZATION

Turning now to Figure 10.6, we see four boards arranged in a panel. Unless you're making only a single board, say as a prototype, you need to arrange them into a pattern. This is for two reasons. First, the PCB manufacturer has standard size copper boards. The smallest is typically 9 in. × 12 in. We're showing here a 2 × 2 panel as an example. (The details are only shown for one board, since the others are exact duplicates.) A real panel would have 8 × 11 boards of this size in the panel to use up the whole sheet of copper stock. To make your little 1 in. diameter board, the manufacturer has to use up a whole board anyway. It is thus more efficient use of

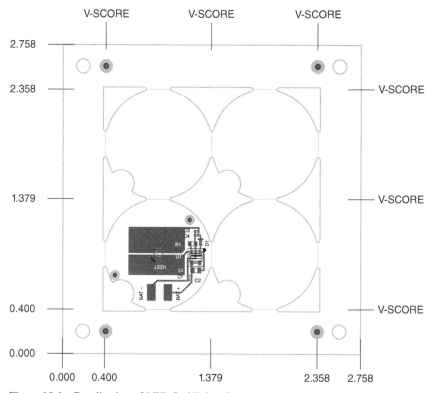

Figure 10.6    Panelization of LED flashlight. (See color plate section.)

material to make a whole panel of the boards at the same time. A second reason for panelization is that when the boards are assembled, the assembly machine has setup time. It's much more efficient for it to stuff a dozen boards (or whatever) at a time. More efficient means cheaper.

The differences between the single board shown in Figure 10.5 and those in Figure 10.6 are a little subtle. The most immediately noticeable is that there is now a semicircular cutout in the board. This is not an electrical thing. But the layout engineer noticed that we have wires coming up to the board, to be soldered onto the BAT pads. Since the PCB is physically tight inside the flashlight's housing, there would be no place to run the wires. So she's put the cutout into the board outline as a place for the wires to go.

Another difference on this panelized version is that there are now four flattened sides in the circle. This is so that the board repeats are attached by about 0.2 in., rather than at just a single point if they were left as circles. This gives more rigidity to the panel, and helps with assembly and handling. You'll also notice that the common flat edges have a "chamfer," or a rounded radius. This allows the cutouts to be routed by a round drill bit. If we had left them as a sharp wedge as in between two circles, the PCB manufacturer would not have been able to produce it.

The other reason for having the flat sides is for depanelization. During assembly, a layer of solder paste is spread on. Then the components are placed and the entire panel is put through a reflow oven. In the oven, the paste is heated to its melting temperature, soldering the components down to the board. The final step of the process is to depanelize (or cut apart) each board. The fastest way to do this is to run a saw straight across the edges. Of course, this works properly only when cutting straight lines. On the panel you will see places where "v-score" is labeled. These are actually prescored when the PCB is manufactured so that the final depanelization is quicker and easier.

The entire panel of 2 × 2 boards is surrounded by a 0.4 in. border. This is called the rail. The rail surrounds all four sides, although sometimes it's just on the two longer sides. You'll notice four 0.12 in. diameter holes for mounting onto the assembly machines, as well as four additional fiducials for machine alignment. The rails are necessary for assembly.

We note that the panelization can be done by PCB manufacturers if the boards are simple rectangles, or when assembly is done by hand. However, if you have round boards or want to make sure the rails and v-scores are optimized for assembly and depanelizing, then it might make sense to do your own panelization.

## FINAL DESIGN

In the end, our design didn't use any electrolytic capacitors. The initial concern about low-temperature operation then comes down to possible moisture condensation on the PCB. We can take care of this by conformal coating of the PCB. The final specification will thus go down to −40 °C, as marketing wanted in the first place. It sounds like a good device. We'll have marketing get us one from the production line.

# CHAPTER 11

# PRACTICAL DESIGN OF A USB LIGHT

## INITIAL MARKETING INPUT

For their next product, marketing wants to come out with a USB reading light. This is apparently inspired by the VP of marketing, who wants to see his computer's keyboard on an airplane without turning on the overhead light. (He doesn't want to use a Mac!) "I want to plug it into the USB port on my laptop, and then have a flexible neck that leads up to a little LED. It doesn't have to be very bright. No, on second thought let's make it as bright as you can, and we'll put a control on it so I can adjust the brightness. I bet lots of people wish they had something like that."

Since there is only so much power a USB port can deliver, that sets the upper end of the brightness. If we use a resistor divider as the control, the intensity will be nonlinear with the control setting, but "I don't think I care about that." How about an on/off switch? "Too much stuff. When the dial is at the bottom of its scale, you can turn it off then." And should it be totally off, or just really dim? "Oh, if you can't see it, it's OK. It can't be much power that way anyway, right? I guess I don't want to drain my computer's battery." We promise to evaluate standby current drain from the USB port. We also need to remember to look at efficiency, since that too will influence battery life and maximum output brightness. Marketing didn't specify the cost of the light, but they will of course want to know that before approving the design.

## INITIAL ANALYSIS

The first thing is to look up the USB specification. Wikipedia[1] informs us that "The USB 1.x and 2.0 specifications provide a 5 V supply on a single wire from which connected USB devices may draw power. The specification provides for no more than 5.25 V and no less than 4.75 V (5 V ± 5%) between the positive and negative bus power lines. For USB 2.0 the voltage supplied by low-powered hub ports is 4.4 V to

---

[1] http://en.wikipedia.org/wiki/Universal_Serial_Bus#Mobile_device_charger_standards under license http://creativecommons.org/licenses/by-sa/3.0/

---

*Practical Lighting Design with LEDs*, Second Edition. Ron Lenk and Carol Lenk.

5.25 V. . . . A unit load is defined as 100 mA in USB 2.0, and was raised to 150 mA in USB 3.0. . . . All devices default as low-power but the device's software may request high-power."

To be backward compatible, then, and to avoid having software, our light can't draw more than 100 mA. And to be compatible with 2.0, the voltage can be as low as 4.4 V. So the maximum input power to the light shouldn't exceed 440 mW. This means that for this design, we again don't need a big LED. A 1 W or even ½ W LED would be enough.

Since we're going to use a single white LED, the output voltage of the driver is always going to be less than the minimum 4.4 V input. Therefore, we need a buck converter. We plan to use the LM3405, since we've already studied it. In addition, its high switching frequency means the components will be small. Come to think of it, could it be made so small as to fit into the USB plug? A quick calculation suggests the power level is compatible with this idea. The datasheet suggests 84% efficiency at 5 V input, that's a loss in the power converter of only 70 mW. That shouldn't be a problem for something the size of a USB plug. Besides, it's plugged in to a USB port, which should offer some heat sinking.

Let's estimate light output. We have an input power of 440 mW, and an assumed efficiency of 84%, so power to the LED will be 370 mW. For this application, we imagine that the CCT should produce comfort for working—not 2700 K, that's too yellow, but not 6500 K either, as that's too stark. We'll use 5000 K LEDs. (Remember to check this assumption with marketing.) With a device at about 150 Lm/W (as we've learned from our last design), we have maximum light output of about 56 lumens. With a little more cost, we can probably do 10–20% more than that. (We'll add in the temperature effects in the more detailed analysis.) Using the estimate from Chapter 3, the illuminance on the keyboard will be about 264 Lux. While this is pretty bright, that might be good for reading printed material. It has a dimmer control anyway.

We're also supposed to figure out the standby power. If we just let the dimmer go down to 10% minimum rather than turning off, the input power will be approximately 44 mW. The assumption for making this approximation is that fixed overhead power draw, such as the IC operating current, is still a small contribution to total power even at this power level. Since 44 mW at 5 V (nominal USB output voltage) is 9 mA, and the IC draws 1.8 mA typically, this looks like an OK approximation. A typical laptop battery has an ampacity of maybe 4000 mAh, so 9 mA current drain would be a 0.2% capacity load. And that's if it was linearly regulated down. We know that in reality the USB port is actually powered by a little SMPS, and that the laptop battery is probably 19 V. So we should compare power, not current: The laptop battery has an energy storage of 4 Ah × 19 V = 63 Wh, and so a 44 mW load would take 1430 h = 2 months to drain the battery. We guess that this level is OK, subject to marketing approval. After all, it will still be noticeable that the light is on, so you can unplug it; and USB ports typically are turned off by the computer when the laptop goes into hibernation anyway.

The last thing is to estimate cost. The LED we will again cost out at $0.10. The LM3405 costs $0.51 ($0.56 with an exposed pad on the bottom for heatsinking) in 1 K pieces. We'll guess the cost of the rest of the components to be $0.30, giving a total of

**TABLE 11.1 Initial Specification for USB Light**

| Parameter | Min. | Typ. | Max. | Units |
|---|---|---|---|---|
| Maximum brightness | | 56 | | Lumens |
| CCT | | 5000 | | K |
| CRI | | 70 | | – |
| Illumination | | 264 | | Lux |
| Dim range | 10 | | 100 | % |
| Input power at maximum brightness | | | 440 | mW |
| Input power at minimum brightness | | 44 | | mW |
| Touch temperature | | | 60 | °C |
| Ambient temperature (by design) | 10 | 25 | 40 | °C |
| MTTF | | 50,000 | | h |
| BOM cost | | | $1.00 | US$ |
| Total cost | | | $3.00 | US$ |

about $1.00. Operations hopes the mechanical parts are of about $1.50, and assembly cost for the PCB and the mechanical parts will be $0.50. Thus the total cost will be about $3. With the same margins we used on the LED flashlight, the retail price would then be about $15.

## SPECIFICATION

We prepare a specification for this design, see Table 11.1.

Marketing expresses some concern about the 10% minimum level. How dim is that going to be, is it going to keep the VP awake? And is 10 mA really the best we can do? We point out that the USB port turns itself off when the computer is put to sleep, but agree to see how low we can go. Perhaps they'd like to turn the thing 100% off when the level is below 10%? This could be done with a comparator on the EN/DIM pin. "If you can do it without affecting the cost or footprint" is the answer. So that doesn't work out.

As for the CCT, they think that it should be more like a reading light, the color is somewhat more important for typing and reading than for a camp flashlight. "What's the penalty for going down to 4000 K?" The answer we give is that there's probably no cost penalty, and the CRI rises from 70 to 80, but the light is decreased 10%. Since it's pretty bright already, they like that better. The agreed-on design spec then is shown in Table 11.2.

## POWER CONVERSION

The very first thing we need to do is to select the LED, because everything else depends on that selection. As in the previous chapter, we're going to walk through component selection, albeit a bit faster this time. Going to Digikey, we look for "LED," then "LED Lighting—White," and "In stock", which leaves us with 2985

**TABLE 11.2  Design Specification for USB Light**

| Parameter | Min. | Typ. | Max. | Units |
|---|---|---|---|---|
| Maximum brightness | | 50 | | Lumens |
| CCT | | 4000 | | K |
| CRI | | 80 | | – |
| Illumination | | 240 | | Lux |
| Dim range | 10 | | 100 | % |
| Input power at maximum brightness | | | 440 | mW |
| Input power at minimum brightness | | 44 | | mW |
| Touch temperature | | | 60 | °C |
| Ambient temperature (by design) | 10 | 25 | 40 | °C |
| MTTF | | 50,000 | | h |
| BOM cost | | | $1.00 | US$ |
| Total cost | | | $3.00 | US$ |

choices. Choosing 4000 K reduces this to 492 choices, and adding in that the forward voltage should be less than 3.8 V (so we can use the buck), 343 choices. We'll also cut out anything less than 100 Lm/W. Although we told marketing it would be about 150 Lm/W, somewhat lower is probably OK if it's cheaper. We don't want to corner ourselves on pricing by being overly aggressive on the performance. So in the end we have 277 choices for the LED.

We now sort by price, low to high. The first device that comes up is the same LED we used on the flashlight, the Luminus MP-2016. But this has a maximum current of 120 mA at a forward voltage of 3 V, so putting 370 mW into it is really too much. (Digikey doesn't offer a selection based on power. We could have sorted by maximum current, but since the forward voltage varies from 2.76 to 3.8 V, this might eliminate some parts that would work.) The next device down the list is 100 mA at 3.1 V, so that doesn't work either. But the third device down has a test current of 150 mA at 3 V, which is fine; an efficacy of 162 Lm/W and a CRI of 80; and costs $0.08 at 5000 pieces. What about the size, is it too big? It's 3 mm × 3 mm, so that seems nice and small. We choose this LED, the Luminus MP-3030-1100-40-80.

Having chosen the LED, now we can begin designing the driver. The first thing is to figure out the drive current. We approximated the LED power as 370 mW, which at 3 V is about 120 mA. From the datasheet, the forward voltage of the LED at this current is about 3.08 V at 25 °C. For our initial design, we'll assume the thermal pad of the LED in steady-state is going to be about 60 °C. The forward voltage looks like it drops by about 40 mV at this temperature. We thus expect a forward voltage of 3.08 V − 40 mV = 3.04 V. Dividing this into the power, we get an LED target design current of 122 mA, very close to our estimate.

Our initial schematic, taken from the datasheet of the LM3405, is shown in Figure 11.1. The only important difference is the addition of the potentiometer $R2$ (or pot, as it's called). (By the way, what's the difference between a potentiometer and a rheostat? As far as we know, the only difference is that a pot is a low-power device for trimming resistance values, while a rheostat is a device for dissipating power.) The pot

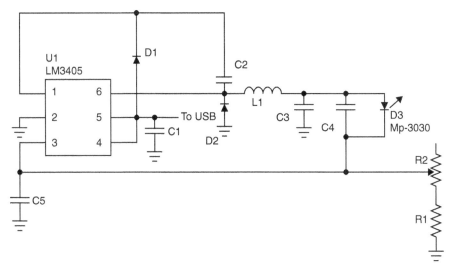

Figure 11.1    First-cut USB light schematic.

is going to be used to do the dimming. The way the LM3405 works is that it controls the current through the LED such that the voltage on pin 3 will be 205 mV. Now when $R2$ is set to zero ohms (all the way to the bottom), the voltage across $R1$ will be 205 mV, and so the current in the LED will be 205 mV/$R1$. As we now increase the value of $R2$, the resistance from the LED to ground increases. We'll now have the same voltage, but a larger resistance. Thus, the current through the LED will decrease, becoming 205 mV/($R1 + R2$). And that's how we dim the LED.

We need to select values for the components L1, C1-5, D1 and D2, and $R1$ and $R2$. Starting with the capacitors, the LM3405 datasheet recommends a 10 μF ceramic for C1, with an X7R or X5R dielectric. Since it's attached to the 5 V input, a 6.3 V rating would be sufficient. Now because of the plan to fit the ballast into the USB plug, we need to be really careful about sizing components. We go through the same procedure for selecting the components as we did in the previous chapter, but now with an emphasis on size over cost—as long as it's not outrageously expensive. We still have to meet our cost target.

As in the last chapter, we go to Digikey to look for capacitors. We select "ceramic capacitors" and "in-stock," giving us 74,630 choices. We want 10 μF, giving 2573 choices, surface mount only, and X7R or X5R temperature coefficients, reducing the choices to 1866. Now, although we said the part needs to be only 6.3 V, we don't want to exclude other, higher voltages that might possibly be cheaper. So we select "Voltage – Rated" to be in the range of 6.3—16 V, leaving finally 1,100 choices. Since all of these will work, we can look at the sizing because of the plan to fit the ballast into the USB plug. The smallest is 0402—astonishing! However, it runs 19 ¢. Let's instead look at 0603. We can get one of these for only 3 ¢, so we'll pick this for the time being. If the design turns out not to fit, we can shrink to the 0402 for an extra 16 ¢.

The boost cap C2 is recommended to be 10 nF, which of course will be available in 0402. In fact, a quick check on Digikey shows it being available down to a 0201! (When reading the datasheet on these value suggestions, remember that we have renumbered the components in our schematic following our plan of consecutive numbering from upper left to lower right. The datasheet didn't do this, so we have to be careful that we pick the right component when looking at their recommendations.)

The datasheet recommends 1 μF in a 1206 package for the output capacitor C3. The same capacitance value is recommended for the compensation capacitor C4, but a 0603 will do since there isn't any power flowing through it. Again, a quick check on Digikey for 1 μF in at least 4 V (since it's across the LED) shows sizes down to 0201, so we go to 0402. (0201 makes it essentially impossible to work on in the lab. We can always switch later if necessary to make it fit.)

Finally, C5 is a noise filter for our resistor divider. Since we expect to be using something on the order of 10–100 Ω for $R1$ and $R2$, we don't want to use the same 1 μF cap, as otherwise the LED might have a visible flash at turn-on while the FB pin voltage is rising. It is reasonable to have 10 nF, which gives a time constant of 1 μs. This too will be a 0402. Summing up, we have three capacitors in 0402, one in a 0603 and one in a 1206. And we have the option of shrinking some of these further if necessary.

Turning now to the diodes, the datasheet says to use a 1N4148 for the boost diode D1. We find this in a SOD323 package, which is 2.5 mm × 1.25 mm and costs about 2 ¢. For D2, the datasheet says to use a schottky. (We need to remember to change the symbol in our schematic.) We could just pick one that is rated for the full output current of the LED, 120 mA. In fact, we can just reuse the 500 mA schottky we used in the flashlight, the CTS05S30, which is in a really tiny package, only 1.0 mm × 0.6 mm.

To select the resistors, let's review how the circuit consisting of $R1$ and $R2$ works. The IC is going to force the current through the LED to be such that the voltage on pin 3 is going to be 205 mV. The resistance from that pin to ground sets what that current is going to be, since $205 \text{ mV} = I_{LED} \times R$. Now when the pot $R2$ is set to zero, the current is going to be $205 \text{ mV}/R1$. As the value of the pot increases, the total resistance to ground from pin 3 increases, so the current goes down, $I_{LED} = 205 \text{ mV}/(R1 + R2)$. So the maximum current occurs when $R2$ is set to zero. So now we know how to calculate $R1$. We want the current to be a maximum of 122 mA, and so $R2 = 205 \text{ mV}/122 \text{ mA} = 1.68 \, \Omega$. There's insignificant power dissipated in the resistor, since $(205 \text{ mV})^2/1.68 \, \Omega = 25 \text{ mW}$, and so we pick a 1.69 Ω resistor in 0402, noting that we could go down to a 0201.

Now we can select the value of R2. We want the dimming to drop down to 10% of full brightness. That would happen if the resistance to ground was 10 times bigger, that is, $R1 + R2 = 16.8 \, \Omega$, which would happen if $R2 = 15.1 \, \Omega$. Presumably, what marketing has in mind is a rotary wheel, not something that needs a pot tweaker! Since we haven't used one of these before, let's walk through the selection process. Typing in "potentiometer" at Digikey, we get a selection of items, and we can choose "Rotary Potentiometers, Rheostats." A quick search on Digikey shows that there actually is a 15 Ω pot—but it costs $23!! A look at realistically priced pots shows that the next value up is 500 Ω, at $0.84 (if we picked a smaller value, it wouldn't dim to 10%). A 1 kΩ is available for 30 ¢.

Is there any other criterion we need, or can we just pick the cheaper one? It turns out that there are three main choices. We can select the "Taper" to be linear, logarithmic, or reverse logarithmic. So, what happens if we pick the most common type, linear? Remember that $R2$ is in series with $R1$, which we just chose to be $1.69\,\Omega$. If the pot is linear, then half way down, it will be half the value, $500\,\Omega$ (for the $1\,k\Omega$ pot, half that for the $500\,\Omega$ pot). What effect does this have on the dimming? Well at half way down, the current is going to be $205\,mV/(500\,\Omega + 1.69\,\Omega) = 510\,\mu A$. That current is so small that the LED won't be visibly on. So, there's going to be a large amount of turning needed by the user before the LED is visible. And all of the dimming is going to be down at the very bottom, making it hard to control.

For this reason, we settle on a logarithmic taper. There are only eight choices in stock from Digikey, and only one of them appears reasonably priced for our application: The PTV09A-2020U-A102 runs 46 ¢ at 1000 pieces. Opening the datasheet and looking at the taper curve, we see that at 50% travel, the pot will have about 15% of its full resistance, that is, $150\,\Omega$. This gives a current of $205\,mV/(150\,\Omega + 1.69\,\Omega) = 1.3\,mA$, or 1% of full brightness. That will certainly be visible, especially in a dark room. Note that we have also satisfied marketing's goal. At $1\,k\Omega$, the current will be only $205\,mV/1\,k\Omega = 205\,\mu A$ (the extra 1 $\Omega$ is negligible). That current is so small that the LED won't be visibly on, and the drain from the battery will be tiny. We'll evaluate how tiny later on, as we promised to do.

The final item to select is the inductor L1. The datasheet recommends $10\,\mu H$ for a 200 mA load. With this value, the datasheet suggests that peak-to-peak ripple current is 240 mA, so that our peak current would be $122\,mA + (240\,mA/2) = 242\,mA$, with minimum current going down to zero. But we are worried. The ripple current is double the current of the LED, which means it's physically large. And if we approximate the triangular ripple current waveform as a sine wave, the RMS value would be $242\,mA/\sqrt{2} = 171\,mA$. The switcher is running at 1.6 MHz, how much skin-effect losses in the wire are there going to be[2]? Since we don't know, to minimize this we'll instead go with the highest inductance value the datasheet suggests, $22\,\mu H$. This gives a 109 mA peak-to-peak ripple, giving a peak current of 177 mA, and 125 mA rms. Since inductors are the trickiest component to specify, we're going to walk through component selection again, as in the previous chapter.

To start with, then, we go to the Digikey front page and type in "inductor." Selecting "Fixed Inductors," "In Stock," and "Surface Mount" gives us 43,397 choices. Now we decided on $22\,\mu H$, which is a standard value. But we don't want to limit our choices too tightly. We can probably work with anything within, say, ±20%, so we select value from 18 to $24\,\mu H$, leaving us with 2088 choices.

Now, as in the choice of inductor for the flashlight, we need to choose both the current rating and the saturation current rating for L1. Recall that the current rating is how much RMS current can be put through the inductor without making it too hot. Typically, the current rating is the RMS current at which the inductor rises 40 °C. That's not an absolute barrier to running at higher RMS current, but it's a good guide.

---

[2] Skin effect is the tendency of high frequency current to travel only on the outer skin of a wire. This increases the effective resistance and thus losses above that for a plain DC current. Refer to *Practical Design of Power Supplies*, by Ron Lenk, Wiley-IEEE Press, 1998, for a discussion of skin effect.

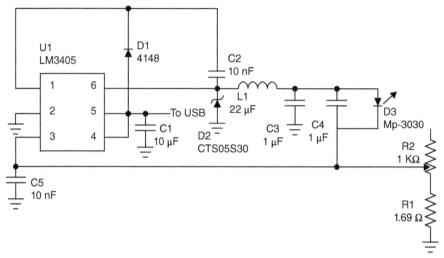

Figure 11.2 LM3405 schematic for USB light.

That's why the manufacturer provides it. Recall further that the saturation current rating is the peak current that can be put through the inductor without saturating the core. Typically, the saturation current is the peak current at which the inductance falls either 10 or 30%, depending on the vendor. Again, that's not an absolute barrier to running at higher peak current, but it's a pretty good guide.

So following our calculation, we need current rating of at least 125 mA rms and a saturation current of at least 177 mA. Again, these are minima. And recall that we estimated the RMS current by approximating the triangle wave by a sine, so the actual RMS current is probably somewhat higher. We thus choose inductors with "current rating" from 125 to 250 mA, and "current—saturation" from 180 to 400 mA. This interestingly leaves us with only nine choices. Our size choices range from 0806 to 1210. Since we're working toward small, we choose the former. Our choice for L1 is then the Taiyo Yuden part CBC2016T220M. This is a 22 µH with a current rating of 240 mA and a saturation current rating of 250 mA. It has a self-resonant frequency (the frequency at which it becomes capacitive) of 16 MHz, so that should be OK. And the cost is only 12 ¢ for a thousand pieces. There's no information on the skin effect, but ignoring that, the losses are about $(0.125A)^2 \times 2.34\,\Omega = 37$ mW, a very modest number. We don't expect it to overheat. The revised schematic is shown in Figure 11.2, with the BOM in Table 11.3. Notice that, again, we did not divide by three to estimate the cost of the IC or the LED. The total material cost came out to be only \$0.95, gratifyingly close to our initial estimate of \$1.00.

## EFFICIENCY

There are two aspects to our efficiency calculations. We are going to start with the power conversion efficiency at full output power. We'll calculate minimum power later. We've assumed in our estimates that the converter will be about 85% efficient.

**TABLE 11.3 BOM for USB Light**

| Ref des | Description | Part # | Mfr. | 1 Kpc price ($) | Est. price ($) |
|---|---|---|---|---|---|
| C1 | 10 µF Ceramic, X5R, 0603, 6.3 V | Any | Any | 0.035 | 0.012 |
| C2 | 10 nF, 0402 | Any | Any | 0.002 | 0.001 |
| C3 | 1 µF, 1206 | Any | Any | 0.031 | 0.010 |
| C4 | 1 µF, 0402 | Any | Any | 0.006 | 0.002 |
| C5 | 10 nF, 0402 | Any | Any | 0.003 | 0.001 |
| D1 | 1N4148, SOD323 | Any | Any | 0.045 | 0.015 |
| D2 | 500 mA, 30 V Schottky | CTS05S30L3FTR-ND | CTS05S30 | 0.066 | 0.022 |
| L1 | 22 µH, 2.34 Ω | CBC2016T220M | Taiyo Yuden | 0.125 | 0.042 |
| R1 | 1.69 Ω, 0402 | Any | Any | 0.004 | 0.001 |
| R2 | 1000 Ω Pot, Rotary, log taper | PTV09A-2020U-A102 | Bourns | 0.456 | 0.152 |
| U1 | LM3405 | LM3405 | National | 0.506 | 0.506 |
| PCB | | | | | 0.079 |
| TOTAL | | | | | 0.843 |
| D3 | LED | MP-3030-1100-40-80 | Luminus | 0.107 | 0.107 |
| TOTAL | | | | | 0.95 |

Now that we have a design, it would be a good point in time to validate that assumption. In order to do that, you'll recall that the efficiency is the output power divided by the input power. And of course, the "output power" is the power into the LED. And so $\eta = $ (power into LED)/[(power into LED) + (power in all the other components)]. The first step, then, will be to calculate the power into the LED.

Now, for some purposes, it would probably be good enough just to multiply the average current, 122 mA, times an approximation of the forward voltage, 3.04 V, and get back to our original power estimate of 370 mW. However, in this case, we probably need to do better. If we pull too much current from the USB port, it might detect a fault. And while this doesn't seem all that likely, we know that the forward voltage increases with increasing current, and the power is thus the product of two things both of which are increasing. So, a little extra engineering time seems worthwhile.

To do this, our plan is to find the current as a function of time, which we can do from the operation of the converter and by using the inductor value. Then, following the ideas in Chapter 15, we're going to find an analytic expression for the *I–V* curve of the LED. We then will use this to express the power as a function of the current, and integrate over one switching cycle to find the energy delivered to the LED. Dividing by the period gives the average LED power (remember that power is energy per time). If it turns out to be too high or too low, we can adjust *R*1 to compensate.

So, the first step is to calculate the *I–V* curve of the MP-3030. We use the datasheet to collect data, and then plot it with Excel, as shown in Figure 11.3 (this is shown in more detail in Chapter 15). Note that, differently from the datasheet, we are here plotting *V* versus *I*, rather than *I* versus *V*. This is so that the curve-fitting produces an expression for voltage as a function of current, not the other way around. When this is multiplied by the current, we get the power, and thus it yields an

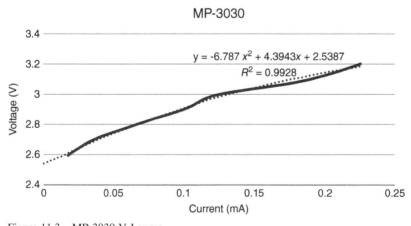

Figure 11.3   MP-3030 V-I curve.

expression for power versus current. We need this because current is what we know from the inductor calculations.

Looking then at Figure 11.3, the solid, wiggly line is Excel drawing interpolating lines between the data points we entered. We took data every 100 mV, and eyeballed the current from the datasheet. We've used Excel's "trend line" function, with the trend line shown as a dotted line. The trend line uses a quadratic fit, and we display both the best fit equation and the correlation coefficient. The latter shows that the fit is close to perfect, and we conclude that the LED forward voltage as a function of current is well expressed as $V = -6.787\,I^2 + 4.3943\,I + 2.5387$, with the current in amps and the voltage in volts. And since the power is this times the current, we have $P = 6.787\,I^3 + 4.3943\,I^2 + 2.5387\,I$, with the power in watts.

Now we need the current into the LED. Since this is a buck converter, we know that the output voltage is a function of the input voltage, and can use that to calculate how long the inductor current is increasing and how long it is decreasing. But the usual formula doesn't take into account the forward voltage of the diode, D2. So to account for that, we're going to do volt-second balance across the inductor.[3]

When the switch (inside the IC) is on, 5 V (from the USB port) is applied to the inductor, while the voltage on the LED is 3.04 V. The LED is in series with the resistor that feeds pin 3, so that's an extra 205 mV. So we have 5.00 V − (3.04 V + 0.2 V) = 1.76 V across the inductor, causing the current to ramp up. When the switch is off, the diode conducts. Its datasheet shows it to have a forward voltage of about 250 mV at 120 mA. Since the diode's anode is at ground, the voltage at the cathode is *negative*. Since the LED is still at approximately 3.04 V, that's (3.04 V + 0.2 V) − (−0.25 V) = 3.49 V across the inductor in the other direction, causing the current to ramp down.

These two numbers allow us to calculate the duty cycle of the inductor (which is, of course, independent of the switching frequency). We have 1.76 V × $t_{on}$ = 3.49 V × $t_{off}$. Adding 3.49 V × $t_{on}$ to both sides gives 5.25 V × $t_{on}$ = 3.49 V × ($t_{off} + t_{on}$) = 3.49 V × $T$, where $T$ is the switching period. And then dividing both sides by $T$ and by 5.25 V, we have the duty cycle DC $t_{on}/T$ = 3.49 V/5.25 V = 66.5%. The current increases 66.5% of the time, and decreases 33.5% of the time. We can also use this number to get the actual ripple current. The period is 1/1.6 MHz = 625 ns. The 1.76 V is applied across the inductor for 66.5% of that time. And so the peak-to-peak ripple current is $I = V \times t/L$ = 1.76 V × (625 ns × 66.5%)/22 μH = 33 mA, substantially less than our estimate of 109 mA.

Now we can finally calculate power in the LED. The current goes from [122 mA − (33 mA/2)] = 105.5 mA to [122 mA + (33 mA/2)] = 138.5 mA 66.5% of the time; and then back down from 138.5 to 105.5 mA the remaining 33.5% of the time. Expressed as equations, the current is

$$I2(t) = 0.1385 - 0.033\,\frac{t}{0.335}, \quad \text{for } t = 0 \text{ to } 0.335.$$

[3] See *Practical Design of Power Supplies*, by Ron Lenk, Wiley-IEEE Press, 1998, for details on why and how we are doing the calculation this way.

And so the power is

$$P = \int_0^{0.665} V \cdot I1 \, dt + \int_0^{0.335} V \cdot I2 \, dt$$

$$= \int_0^{0.665} \left[ 6.767 \left( 0.1055 + 0.033 \frac{t}{0.665} \right)^3 + 4.3943 \left( 0.1055 + 0.033 \frac{t}{0.665} \right)^2 \right.$$

$$\left. + 2.5387 \left( 0.1055 + 0.033 \frac{t}{0.665} \right) \right] dt + \int_0^{0.335} \left[ 6.767 \left( 0.1385 - 0.033 \frac{t}{0.335} \right)^3 \right.$$

$$\left. + 4.3943 \left( 0.1385 - 0.033 \frac{t}{0.335} \right)^2 + 2.5387 \left( 0.1385 - 0.033 \frac{t}{0.335} \right) \right] dt.$$

Evaluating using a computer algebra program, we have $P = 388$ mW. (It could of course be done by hand, but it's kind of messy.) This is pretty close to our estimate of 370 mW—but nonetheless 5% higher. The difference is probably due to the higher forward voltage at the highest current contributing more to the power dissipation than our estimate. At any rate, we'll next calculate the losses, and see if the total is within our power budget.

We're now going to try to estimate power losses in each of the other components, starting from the easiest to calculate or estimate to ending with the most difficult. For this particular design, the easiest components to calculate power loss are the capacitors. If they had been aluminum electrolytic, then we would have had to have looked at their ESR (Equivalent Series Resistance) and calculate their RMS current. We might even have had to look at the temperature coefficient of the ESR. But since our design is high frequency, all of the capacitors are ceramic, and an excellent approximation to their dissipation is zero.

With that out of the way, let's take a look at the power dissipation in the resistors. $R1$ is $1.69\,\Omega$, and has the LED RMS current going through it. Since our power estimate for the LED turned out to be close to our exact calculation, it will be acceptable to use the average LED current as an estimate for this calculation. We have $P = (122\,\text{mA})^2 \times 1.69\,\Omega = 25\,\text{mW}$. As for $R2$, by assumption we are working at maximum output power, which happens when $R2$ is zero ohms. So the power dissipation in $R2$ in this case is zero.

Next we can do the diodes. D1 is the boost diode. Now, to calculate the power dissipation, we should find the current in D1, look up the forward voltage, and multiply. Of course, it is only conducting current while it is charging the boost cap, C2. A careful look at the LM3405 datasheet shows that the switch turns off in 12 ns. So this is how much time D1 is conducting. But we also know how much *charge* D1 delivers. It charges C2, which is 10 nF, up to 5 V. So D1 delivers $Q = C \times V = 10\,\text{nF} \times 5\,\text{V} = 50$ nC of charge. Since it does that in 12 ns, the current through D1 is $I = Q/t = 50\,\text{nC}/12\,\text{ns} = 4\text{A}!$

Looking at the datasheet for the 4148, there is (remarkably for a 200 mA device) forward voltage shown for currents up to 800 mA, at which current the forward voltage is about 1.45 V. It also looks to be increasing geometrically—this makes sense, the diode $I$–$V$ curve is well known to be exponential. It's 1.2 V at 400 mA, so it's increasing 250 mV for each doubling in current. To get to 4 A, we can double three times from 500 mA. At 500 mA, the forward voltage from the datasheet is 1.3 V, so at 1 A it's 1.55 V, at 2 A it's 1.8 V, and at 4 A it's 2.05 V. Its instantaneous power is then 2.05 V × 4 A = 8.2 W (oh-oh!). But this happens with a duty cycle of DC = 12 ns × 1.6 MHz = 1.9%, so the average power in D1 is 8.2 W × 1.9% = 156 mW.

This power really seems excessive; it's nearly half what the LED is getting! What would happen if we just used the same schottky as we're already using for D2? It shows 260 mV at 100 mA and 300 mV at 200 mA, so it increases only 40 mV per doubling. It shows 380 mV at 500 mA, so at 1 A it's 420 mV, at 2 A it's 460 mV, and at 4 A it's 500 mV. Its instantaneous power is then 500 mV × 4 A = 2 W, and so its average power is 2 W × 1.9% = 38 mW, which seems much more reasonable.

D2 is the catch diode. It conducts 33.5% of the time, and its current is the inductor current, an average of 122 mA. Its forward voltage is thus about 265 mV, giving an instantaneous power dissipation of 32 mW. And so its average power dissipation is 32 mW × 33.5% = 11 mW.

The next component for us to calculate the power loss of is the inductor. Inductor loss calculations are tricky. There are two separate sources of power loss in the inductor. One is the $I^2R$ loss in the wire. This is complicated by the high switching frequency, which causes an increase in the effective resistance due to skin effect. The second type of power loss in an inductor is core loss. As the current increases and decreases, the flux in the magnetic core also increases and decreases, and this in turn causes losses in the core material.

We don't know the size of the wire used inside the CBC2016T220, and so we can't calculate how much, if any, the skin effect affects the equivalent resistance. However, we know that the ripple current is only 33 mApp, while the DC current is 122 mA. So most of the current is DC, and the component at 1.6 MHz is small. So we're going to take our best guess, and say that the DC resistance dominates, and that our former estimate of $(0.122 \text{ A})^2 \times 2.34\,\Omega = 35$ mW is about right for the $I^2R$ loss in the wire.

To calculate magnetic core loss, we would have to know the type of ferrite core being used, and also how much (in the sense of volume) core material was being used. Complicating yet further, core losses are always shown for sinusoidal waveforms. In our case, we have a triangle wave. Presumably, we would have to take the Fourier transform of the triangle wave, and look at several of the harmonics to estimate the core losses at each frequency. Since we don't know the core material or volume, and we certainly don't want to get bogged down on complex calculations, we're instead going to do another best guess. It's the same approximation we used for the wire. Since ripple current is small compared to DC, we going to approximate the core loss as zero.

The final component to calculate power loss for is the IC. In this case we're lucky, the datasheet provides a detailed guide to calculating loss. There are two main sources of loss, the operating current of the IC and the losses of the power MOSFET inside the IC. The operating power is easy. The IC draws 1.8 mA and it's powered by the 5 V from the USB port, so that's 9 mW. The MOSFET losses have three components, conduction loss ($I^2R$), switching loss as it turns on and off, and current that turns the MOSFET gate on and off. This last is not usually significant, but in this case the switching frequency is very high and so the gate current, which is gate charge times frequency, is comparatively high.

For conduction loss, the datasheet suggests the approximation $P = I^2 \times R \times DC$ $= (0.122\text{A})^2 \times 0.3\,\Omega \times 0.335 = 1\,\text{mW}$. For switching loss, the power depends on the turn on and turn off times, and how often they are turned on and off. $P = \frac{1}{2} \times V \times I \times f \times (t_{\text{rise}} + t_{\text{fall}}) = 0.5 \times 5\,\text{V} \times 122\,\text{mA} \times 1.6\,\text{MHz} \times (18\,\text{ns} + 12\,\text{ns}) = 15\,\text{mW}$. And for gate charge, $P = f \times V \times Q = 1.6\,\text{MHz} \times 5\,\text{V} \times 1.4\,\text{nC} = 11\,\text{mW}$. Total power loss in the LM3405, then, is the sum of the four contributions, $P = 9\,\text{mW} + 1\,\text{mW} + 15\,\text{mW} + 11\,\text{mW} = 36\,\text{mW}$.

Having (finally!) collected all of the power losses in the system, we can tote everything up, and find the efficiency. The LED power is 388 mW. The power loss by everything else is $P = P_{\text{R1}} + P_{\text{D1}} + P_{\text{D2}} + P_{\text{L1}} + P_{\text{U1}} = 25\,\text{mW} + 38\,\text{mW} + 11\,\text{mW} + 37\,\text{mW} + 35\,\text{mW} = 146\,\text{mW}$, although this might be a bit conservative, given our assumptions. Total power, then is $388\,\text{mW} + 146\,\text{mW} = 534\,\text{mW}$ and efficiency is 388 mW/534 mW = 73%.

Now what do we do? Input power is too high, at 5 V input, 534 mW is greater than 100 mA. Looking at the various terms that factor into the power, there are three categories. There are losses dependent on the square of the current (R1, L1 and conduction loss in the IC), losses linear in the current (D2 and switching loss in the IC), and constant losses (D1, gate charge and IC supply current). Total quadratic losses are $25\,\text{mW} + 37\,\text{mW} + 1\,\text{mW} = 63\,\text{mW}$; linear losses are $11\,\text{mW} + 15\,\text{mW} = 26\,\text{mW}$; and losses that are fixed are $37\,\text{mW} + 11\,\text{mW} + 9\,\text{mW} = 57\,\text{mW}$.

Then, suppose that we cut the LED current by a factor $X$. The quadratic power would divide by $X^2$, the linear power divides by $X$, and the constant losses are, well, constant. LED power also gets cut by $X$. Total power into the circuit would then be $63/X^2 + (388 + 26)/X + 57$, in mW. And we want this to be no more than 440 mW. We can solve this quadratic equation, and the answer is that $X = 1.22$. We thus will decrease the current by this factor. Instead of the 122 mA we had been using, we will use 100 mA, and R1 changes to 2.05 $\Omega$. Just as a matter of interest, let's check to see the efficiency. Substituting in $X$, the LED power is now 318 mW; power losses are 122 mW, and so efficiency is 318 mW/(318 mW + 122 mW) = 72%. This is considerably less than the datasheet led us to suppose, but since total power drain is right, that's just the way it is. We need to revise the datasheet at the end to reflect 20% less light output.

We have yet another calculation to do: standby power. We promised marketing to find out how much power was being dissipated when the control was set to minimum. Fortunately, this is now straightforward. The current through the LED is 205 mV/1 kΩ = 205 μA, so $X = 100\,\text{mA}/205\,\mu\text{A} = 500$. Substituting, we find that everything except the constant IC power loss is zero, and so total power draw is 58 mW. That's quite close to our original guess of 44 mW, and so marketing's requests will be met.

# THERMAL MODEL

We need to estimate the temperature of the LED to make sure we meet the lifetime, and also so we don't burn it up. Both depend on the junction temperature of the LED, so that's what we'll compute. Now marketing has decided to have a slick design for this light. The whole thing is going to be encased in plastic, and they're putting a diffusion lens over the LED. The diffuser is going to lose about 10% of the light. This will reduce our brightness and illumination numbers by 10%, but since this was very bright already and the optical numbers were only typicals, we'll just change the datasheet rather than changing the design.

More significantly for the design, encasing the whole thing in plastic gives the LED very high thermal resistance to the ambient. In the flashlight case in Chapter 10, there was aluminum to attach the LED to; plastic has about 1/500th the thermal conductivity that aluminum has.

So in this case, it seems, most of the heat has to be transferred by the power wires down to the USB plug, where it will be dissipated into the metal case. Let's see how well we can do with these two wires. The distance from the LED to the output of the power supply is about 12 in. From our calculation in the flashlight example, we know that two strands of AWG20 wire of 2 in. have a thermal resistance of 125 K/W, so the 12 in. wire will have 750 K/W. With LED power dissipation of 320 mW, this gives a temperature rise of 240°C! We clearly want to cut this down by at least a factor of 5, and 10 would be even better. How thick does the wire have to be? If we go from AWG20 to AWG18, the diameter increases from 0.032 in. to 0.0403 in., a ratio of 1.259. The area ratio is the square of this, 1.586. The thermal resistance drops by this factor, as does the temperature rise, leaving us with a rise of 151 °C, which is still too high.

Now, wire bigger than AWG18 becomes hard to handle in manufacturing. Besides, it might make the neck of the light stiff. What would happen if we used two strands of AWG20 wire each for plus and minus? With double the area, we get half the temperature, 120 °C. This is still too high. How about using three wires? Now the temperature rise will be 80 °C. At nominal 25 °C, the LED will be sitting at 105 °C, and with a maximum ambient of 40 °C it will be at 120 °C. The datasheet shows that maximum junction temperature for the LED is 125 °C. How much does the junction rise above the case? Power dissipation in the LED was 318 mW, and the datasheet shows that the thermal resistance from junction to case is 11 °C/W, so there's a further temperature rise of 3½° less than 125 °C, so we're OK.

So, the maximum temperature is OK. As for the lifetime, rated maximum current in the LED is 240 mA. We're at only 100 mA, which is less than half, so the lifetime will be better than LM80, which is good enough. The real problem is that at this high temperature, the light output has dropped to probably 80% of nominal. We have no room to maneuver. We can't increase the power dissipated as that will send the temperature over the top. We again have to change the datasheet. But wait! It isn't *that* bad. We just calculated the junction temperature at 40 °C ambient. But what matters for light output is performance at 25 °C, and the case temperature not the junction temperature. Fifteen degrees less takes us to 105 °C, and at 105 °C, light is down about 85%, rather than 80%. We'll change the datasheet accordingly. Touch temperature will be unaffected by the hot LED because the plastic acts as a great thermal insulator—that's why we have thermal problems in the first place.

## MTTF

Looking back at the specifications, the one thing we haven't checked yet is MTTF, mean-time-to-failure. What does that mean? And why isn't MTBF specified? What we're talking about is how long a unit is going to last in operation by the customer. Generally, failures occur with a rate described colloquially as a "bathtub curve." It's high at the beginning, then low for a long time, and then high again towards the end. The initial high failure rate ("infant mortality"—we're not making this up, that's really what it's called) is due to things that are poorly designed, or poorly assembled, or things that are not working quite right, but not so bad that they cause a production test failure. Normally, these sorts of failures are low.

Toward the end of life, units start failing because things wear out. Electrolytic capacitors lose their capacitance, electro-migration occurs on the traces inside the ICs, and so on. It's that part of the curve we're interested in. And specifically, we want to know when half of the units have failed. That's MTTF, the calculation of probably when will half the units stop working. What then is MTBF? That's the way most people refer to this. MTBF stands for mean-time-between-failures. It's a military concept, used for devices that are going to be repaired when they fail. You want to know how often they need to be serviced so that they won't fail while in use. Since this is a consumer device, it's never going to be repaired. If it fails, it'll be thrown away. So what we want is MTTF, not MTBF.

There are a couple of choices to calculate MTTF. For most power supplies, the most common method is MIL-HDBK-217F. It's a military guide put together in the 1990s. It doesn't cover all the device types you might want. And it shows its age in several ways. But there really isn't anything better. So you end up making a number of approximations. The goal of this section is to walk you through how to do these calculations and approximations.

So the first thing is to be clear what we're calculating and how. MTTF is measured in hours. And the MTTF of each component is also measured in hours. Do the hours of each component simply add up to make the total? A moment's reflection shows that that doesn't make sense. That would mean that the more components you have, the longer the life! What is actually done is that all the lifetimes are "in parallel," as though they were resistors. That way, if one component fails way more quickly than all the others, it determines the lifetime by itself, the same as a low resistance in parallel with a number of large resistances dominates the total resistance. So we take the inverse of each component's MTTF, add all of them up, and then take the inverse of that sum. One more thing to note. The lifetime of components depends quite strongly (in fact, exponentially) on temperature. So the high operating temperature is going to seriously impact our MTTF (Figure 11.4).

There are a number of specialty terms used in these calculations. But rather than trying to enumerate all of them, let's just start with a calculation. We'll start with the

| | A | B | C | D | E | F | G | H | I | J | K | L | M | N | O | P |
|---|---|---|---|---|---|---|---|---|---|---|---|---|---|---|---|---|
| 1 | Ref. Des. | Section | $\lambda$b | $\pi$CV | $\pi$Q | $\pi$E | $\pi$T | $\pi$S | $\pi$C | $\pi$R | $\pi$TAPS | $\pi$V | C1 | C2 | $\pi$L | $\lambda$P |
| 2 | C1 | 10.10 | 0.15 | 2.3 | 10 | 1 | | | | | | | | | | 3.45 |
| 3 | C2 | 10.10 | 0.0012 | 1.1 | 10 | 1 | | | | | | | | | | 0.01 |
| 4 | C3 | 10.10 | 0.0012 | 1.9 | 10 | 1 | | | | | | | | | | 0.02 |
| 5 | C4 | 10.10 | 0.0043 | 1.9 | 10 | 1 | | | | | | | | | | 0.08 |
| 6 | C5 | 10.10 | 0 | 1.1 | 10 | 1 | | | | | | | | | | 0.00 |
| 7 | D1 | 6.1 | 0.001 | | 8 | 1 | 9 | 0.054 | 2 | | | | | | | 0.01 |
| 8 | D2 | 6.2 | 0.003 | | 8 | 1 | 9 | 0.054 | 2 | | | | | | | 0.02 |
| 9 | L1 | 11.2 | 0.0057 | | 20 | 1 | | | 1 | | | | | | | 0.11 |
| 10 | R1 | 9.2 | 0.0019 | | 15 | 1 | | | | 1 | | | | | | 0.03 |
| 11 | R2 | 9.14 | 0.1 | | 5 | 1 | | | | 1 | 1 | 1 | | | | 0.50 |
| 12 | V1 | | | | 4 | 0.5 | 28 | | | | | | 0.04 | 0.002 | 1 | 4.48 |
| 13 | | | | | | | | | | | | | | | | |
| 14 | | | | | | | | | | | | | | | Total | 8.73 |
| 15 | | | | | | | | | | | | | | | Hours | 114.609 |
| 16 | | | | | | | | | | | | | | | | |
| 17 | LED | | | | | | | | | | | | | | | 7.70 |
| 18 | | | | | | | | | | | | | | | TOTAL | 16.43 |
| 19 | | | | | | | | | | | | | | | Hours | 60.882 |

Figure 11.4   MIL-HDBK-217F calculation for USB light.

capacitors, all of which in this design are ceramic. Looking at the handbook (it's available online), we see that Section 10 is for capacitors. In Section 10, there are two sections for ceramic capacitors, Section 10.10 for general-purpose ceramics and Section 10.11 for temperature compensating ceramics. Since these are general-purpose caps, we will be using Section 10.10. We enter this information into the first line of our MTTF table, shown as Table 11.4. There are multiple options for which section to use, since we're not using military parts. And so we need to keep track of which section we actually use in case we want to refer back to it.

Starting with C1, we have a 10 μF, 6.3 V cap. We see that the failure rate, $\lambda_p$, depends on the product of four values, $\lambda_b$, $\pi_{CV}$, $\pi_Q$, and $\pi_E$, for each of which we create its own column. Starting with $\lambda_b$, there are three tables available: one for 85 °C parts, one for 125 °C parts, and one for 150 °C parts. Since we see right away that the first table goes only to 80 °C, we're going to need to specify 125 °C parts. Going back to Digikey, the price for the 85 °C part was 2.5 ¢, but for the 125 °C part it's 11.6 ¢! So the hot running temperature of this board is going to impact our cost.

The 125 °C table shows that the failure rate also depends on the stress, which for a capacitor is the voltage it sits at, divided by its rated voltage. In our case, C1 is at 5 V, and it's rated at 6.3 V, so that's a stress of 80%. Looking at the table, we see listed 100 °C and 110 °C, and stress of 0.7 and 0.9. While there's a formula below the table, we'll simply take a guess that it's between those four numbers, and set $\lambda_b$ to 0.15.

The second factor is $\pi_{CV}$, a factor dependent on the value of the capacitor. The highest value is for 4.7 μF, which is 0.3 higher than for 1.1 μF. So we'll guess 2.3. $\pi_Q$ is the "quality factor," which for military designs means how much testing it undergoes. Since we're using commercial parts, we'll always use the row "Lower", in this case 10. Finally, we have $\pi_E$, which is the environment the part operates in. Again, this is a commercial design, so we'll always be using the lowest fact ($G_B$, ground benign), which is 1.

The product of the four terms is 3.45. This corresponds to an MTTF of 1/ 3.45 = 0.29 million h, 290,000 h. So this is probably OK. We can now go through and fill out the table for all the other capacitors. Since they are all similar, the calculations are similar. We just have to remember to select 125 °C parts for all of them. C2 is 10 nF in a 0402, and has 5 V on it. The cheapest part turns out to be 25 V, so we select that. That gives a stress of 5 V/25 V = 0.2. C3 is 1 μF in a 1206, and has 3 V on it. The

**TABLE 11.4  Final BOM for USB Light**

| Ref Des | Description | Part # | Mfr. | 1Kpc price ($) | Est. price ($) |
|---|---|---|---|---|---|
| C1 | 10 μF, X5R, 0603, 6.3 V, 125 °C | Any | Any | 0.116 | 0.039 |
| C2 | 10 nF, 0402, 25 V, 125 °C | Any | Any | 0.002 | 0.001 |
| C3 | 1 μF, 1206, 16 V, 125 °C | Any | Any | 0.031 | 0.010 |
| C4 | 1 μF, 0402, 6.3 V, 125 °C | Any | Any | 0.019 | 0.006 |
| C5 | 10 nF, 0402, 25 V, 125 °C | Any | Any | 0.002 | 0.001 |
| D1 | 500 mA, 30 V Schottky | CTS05S30L3FTR-ND | CTS05S30 | 0.066 | 0.022 |
| D2 | 500 mA, 30 V Schottky | CTS05S30L3FTR-ND | CTS05S30 | 0.066 | 0.022 |
| L1 | 22 μH, 2.34 Ω, 105 °C | CBC2016T220M | Taiyo Yuden | 0.125 | 0.042 |
| R1 | 2.05 Ω, 0402 | Any | Any | 0.004 | 0.001 |
| R2 | 1000 Ω Pot, Rotary, log taper | PTV09A-2020U-A102 | Bourns | 0.456 | 0.152 |
| U1 | LM3405 | LM3405 | National | 0.506 | 0.506 |
| PCB | | | | | 0.079 |
| TOTAL | | | | | 0.881 |
| D3 | LED | MP-3030-1100-40-80 | Luminus | 0.107 | 0.107 |
| TOTAL | | | | | 0.99 |

cheapest part turns out to be 16 V, so its stress is 0.2. C4 is again a 1 µF, this time in a 0402 and also has 3 V on it. The cheapest is 6.3 V, so the stress is 0.5. Finally, C5 is again 10 nF and has only 0.2 V on it. Using the same cap as C2, the stress is essentially zero. It should last forever (at least compared with everything else). At the end of the calculations of the capacitor MTTFs, we see that C1 dominates all the other capacitors. If we need to increase the lifetime, we'll take a look at increasing the size of C1, which will help.

Diodes are next. Discrete semiconductors are listed in Section 6, and nothing we are doing is RF or microwave, so we use Section 6.1. $\lambda_p$ now depends on the product of six values. $\lambda_b$, $\pi_Q$, and $\pi_E$ are still there, but there are also three new ones, $\pi_T$, $\pi_s$, and $\pi_C$. We add these to new columns in our table. Starting then with $\lambda_b$, we decide that the boost diode is probably a switching device, rather than general purpose analog or power. This gives $\lambda_b = 0.001$. $\pi_T$ for a switching device at 105 °C is 9.0. Reverse voltage applied to the diode is 5 V, and it's rated at 75 V, so stress is 0.07. The handbook says that for all stress less than 0.30, the stress factor is $\pi_S = 0.054$. The construction is certainly not metallurgical bonded, so $\pi_C$ is 2. And finally, we're in a plastic case and ground benign conditions for $\pi_Q$ and $\pi_E$. The MTTF for this diode is thus much less than that of the 10 µF C1.

D2 follows the same plan. However, it's a schottky diode, and we decide that it is carrying power. So $\lambda_b$ is triple what it was for D1. The only thing we need to evaluate is the stress. The diode sees 5 V when the IC switch is on, and it's rated at 30 V, so stress is 0.16. Since this is still less than 0.30, $\pi_S$ is the same.

The inductor L1 is called a "coil", since it's not a transformer, and we use Section 11.2. This time we already have columns for the four factors entering into the calculation. Three of them are easy. It's a fixed value inductor, it's not a military standard quality, and it is ground benign. The tricky part is the base failure rate, $\lambda_b$. Now what they want you to do is to measure the "hot spot" temperature. What this really means is sticking a thermocouple inside the inductor while it's running. We're not building this inductor, we're buying it, so there's no convenient way to do that. But we can take an estimate. Remember that we're running this inductor at 100 mA. But it's rated for 240 mA, and so wire losses ($I^2R$) are going to be (100 mA/ 240 mA)$^2 = 17\%$ of rated. So we're going to assume that the operating temperature of the inductor is basically the same as the ambient, 105 °C. And since the part is also *rated* at 105 °C, we find that $\lambda_b = 0.0057$. This gives a $\lambda_p$ of 0.11, still a lot less than C1.

Next comes R1. Resistors are in Section 9. And it's a film resistor, and doesn't carry power, so we use Section 9.2. The lifetime depends on four factors, of which $\pi_R$ is the only new one. Now there are two possibilities for $\lambda_b$ and wehave not really either one. But since their values aren't very different, we'll just use the lower temperature table. Stress for a resistor is power dissipated divided by rated power. Power dissipated is 205 mV$^2$/2.05 Ω $= 21$ mW and the 0402 resistor is rated at 1/16 W, so the stress is 0.33, giving $\lambda_b$ of 0.0019. Resistance is less than 100 kΩ, so $\pi_R$ is 1, quality factor is "lower" and we're ground benign. The $\lambda_p$ ends up being 0.03.

R2 is different because it's a potentiometer. That falls under the class "variable." The datasheet for the part says the resistive material is "carbon," so we'll use Section 9.14, "Variable, Composition." $\lambda_p$ depends on six variables, with $\pi_{TAPS}$ and $\pi_V$ new. Now at full power, the operating power in R2 is zero—because full power occurs when the pot is set to zero.

Where does maximum power occur? The voltage across the whole string, R1 and R2 together, is 205 mV. So the current through the string is 205 mV/(R2 + 2.05 $\Omega$). But then the power in R2 is this current squared times R2, $P = (205 \text{ mV})^2 \times R2/(R2 + 2.05 \, \Omega)^2$. A quick piece of calculus (take the derivative with respect to R2, and set it equal to zero) shows that (unsurprisingly) this is maximum when R2 is also 2.05 $\Omega$, at which the power in R2 is 5 mW. R2 is rated at 50 mW, so the stress is 0.1, and $\lambda_b$ is 0.1. The pot is less than 50 k$\Omega$ so $\pi_R$ is 1, the quality factor is nonmilitary and we're ground benign.

This leaves $\pi_{TAPS}$ and $\pi_V$. We have three taps on this pot: the two at the ends, and the one in the middle that gets adjusted, so $\pi_{TAPS} = 1$. The maximum voltage this pot could ever see is 205 mV, so the voltage factor is surely less than 0.8, and $\pi_V = 1$. We end up with $\lambda_p = 0.5$—much higher than R1, but still only the second largest contributor to MTTF, following C1.

The final component is the IC, the LM3405. This is the fun one. Microcircuits are in Section 5. While power supply controllers—not to mention ones with power devices inside—weren't even invented yet when this handbook came out, the right section to use is 5.1, "Gate/Logic Arrays and Microprocessors." Well, the others fit even worse. And Section 5.1 does call out "MOS devices," which the LM3405 must be, given its switching frequency.

There are three new parameters, $C_1$, $C_2$, and $\pi_L$, and it's no longer a straight product. $C_1$ is multiplied by $\pi_T$, and $C_2$ by $\pi_E$, and then those two products are added together as one of the three factors being multiplied together. Let's start with $C_1$. We already decided it must be MOS, not bipolar. And it's presumably linear, not digital. But we need to know how many transistors it has inside. Now sometimes, the datasheet has this number. But not this time. However, it does have an internal block diagram. This shows a lot of functions, so it's certainly not in the "1–100" transistor category. It might be "101–300" transistors, but to play it safe, we'll say it's "301–1000" transistors. It seems really unlikely to have more than 1000 transistors in the design, and so $C_1 = 0.04$.

Now the short remainder of Section 5.1 tells us that we have to look at *other* sections for the other parameters' values. We turn to Section 5.8 for $\pi_T$. We have "linear (Bipolar and MOS)," which seems right. We now need the temperature of the IC's junction. We decided that at full power, the IC is dissipating 36 mW. The LM3405 package without the heat slug on the bottom is 118 °C/W thermal resistance to the ambient, so the junction temperature will be 105 °C + 0.036 W × 118 °C/ W = 109 °C. (We see that adding the slug in could at most reduce the temperature 4 °C, and so it's not worth the extra cost.) Going to our table, $\pi_T$ is 28.

We turn to Section 5.9 for the $C_2$ factor. Our package is non-hermetic and has five pins, so $C_2 = 0.002$. And then turning to Section 5.10, $\pi_E = 0.5$. $\pi_Q$, the quality factor is a bit challenging. It lists only military-screened parts. The highest value is 2, so we'll multiply that by 2 to be safe. The final value is $\pi_L$, the "learning factor," which is how many years the IC has been in production. The thought here is that if it's less than 2 years, the bugs perhaps have not all been worked out, so it's less reliable. The datasheet, however, shows the original publication year of 2006, so the IC is more than 2 years old and $\pi_L = 1$.

The lifetime for the IC ended up $\lambda_B = 4.48$, just ahead of C1. Adding up all the numbers we get 8.73, which is $1/8.73 = 0.115$ million h, so MTTF = 115,000 h, a quite respectable number for a power supply. The last thing we need is the life of the LED. We're not going to use the handbook for this, because LM80 (with TM-21) provides testing of this directly. To get the test report usually requires contacting the vendor. In this case, we have access to a similar part from the same vendor. It shows projected lifetime of the LED as >54,000 h, even at 105 °C. Now we're running the LED at 100 mA, but it's rated at 240 mA. Since the power dissipation is linear in the current for an LED, we'll estimate the lifetime as 240 mA/100 mA × 54,000 h = 130,000 h.

The final lifetime of the design including the LED is shown in Table 11.4. We add the inverse of the lifetime of the driver (8.73) to the inverse of the lifetime of the LED (7.70), and take the inverse, showing that this device has an MTTF of 61,000 h. This meets our spec of 50,000 h, although we had to switch to 125 °C capacitors.

At this point, you're probably thinking "some of those calculations seem pretty nebulous. Can this really be trusted?" The answer may not please you. You can claim to your customer that you have an MTTF of 61,000 h. They could compare this with other vendors who have shorter lifetimes—if their calculations were done correctly. More commonly, only the driver lifetime is reported, and here 130,000 h is excellent compared to others. One thing the analysis *did* bring to light is that there is an avoidable significant failure mechanism. If you want to make this design more robust, you must choose a different capacitor for C1. Ultimately, the MTTF can only be determined directly, by measurement. But in the end, the estimate we've made here probably isn't wildly off. And it took a few hours, rather than a couple of months of testing.

## PCB

We have the finalized BOM shown in Table 11.4, including the change to D1 and the changes required for our MTTF calculation. In particular, the capacitors are now all rated at 125 °C, and the inductor states that it is 105 °C. That way, when it comes time to transfer the design to manufacturing in China, they (hopefully) won't substitute "equivalent" parts that are actually lower temperature.

We pass the schematic off to layout, getting back Figure 11.5. The resistor numbers have been swapped, with R1 becoming R2 and vice versa. This is to comply with out conventions of numbering from upper left to lower right.

This PCB is considerably more space constrained than the previous one. The USB dongle is not very large in the first place. Some of it is taken up with the connector, and the potentiometer is fairly big. See Figure 11.6, which shows the PCB. The outer two pins of the USB connector are 5 V and ground, and the inner two are not used for this design. (They are for data.) C1, the input bypass capacitor, is right next to the IC. The LED return connection (marked "K" for cathode) has a bit of a run to R1 and R2, the current sense resistors, but there's no avoiding that with a big pot.

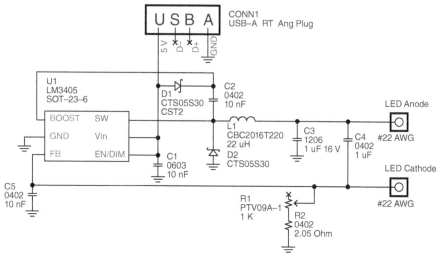

Figure 11.5 Final LM3405 schematic for USB light.

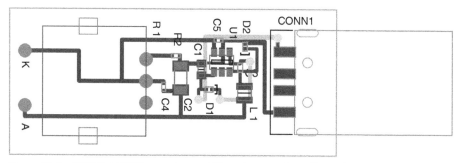

Figure 11.6 USB light PCB, top view and X-ray bottom view. (See color plate section.)

Presumably the board is so tiny that it will be fine. The presence of C5 right next to U1 forms a good filter for the IC inputs. Note that the ground side of R2 joins up with the ground side of C3, the output capacitor, and from there goes right up to pin 2 of the IC to avoid offset voltages. For this design, since the current is very low and the switching frequency very high, we have concentrated on getting parts close together rather than on having a ground plane. The only other thing to notice is that the USB connector hangs out of the USB dongle slightly, allowing it to be plugged in.

## FINAL DESIGN

Our final specification is shown in Table 11.5. Maximum flux was originally 50 Lm; we divided that by 1.22 to keep the input power in spec, and then we dropped it a further 15% because of its operating temperature. To double check, we go back to the datasheet. The MP-3030-1100-40-80 has a typical flux of 67 Lm at 150 mA at 25 °C.

**TABLE 11.5    Final Specification for USB Light**

| Parameter | Min. | Typ. | Max. | Units |
|---|---|---|---|---|
| Maximum brightness | | 40 | | Lumens |
| CCT | | 4000 | | K |
| CRI | | 85 | | – |
| Illumination | | 192 | | Lux |
| Dim range | <1% | | 100 | % |
| Input power at maximum brightness | | | 440 | mW |
| Input power at minimum brightness | | 60 | | mW |
| Touch temperature | | | 60 | °C |
| Ambient temperature (by design) | 10 | 25 | 40 | °C |
| MTTF | | 60,000 | | h |
| BOM cost | | | $0.98 | US$ |
| Total cost (target) | | | $2.98 | US$ |

At 100 mA, that drops to 72% of that number. And at 105 °C, it looks like a further drop of 83%. So total flux will actually be 40 Lm.

Dim range now goes extremely low, since the 1 kΩ pot is huge compared with the 2 Ω sense resistor. Rather than saying 0%, which would imply there's an off switch, we list minimum brightness as less than 1%. Minimum input power per our calculations is listed as 60 mW. We increase the MTTF to 60,000 h, and change our cost to our final estimates. Looks like we've met all of our specifications!

# CHAPTER 12

# PRACTICAL DESIGN OF AN AUTOMOTIVE TAIL LIGHT

## INITIAL MARKETING INPUT

Our customer is an after-market provider of automobile lights. They come in with a pretty good idea of want they want. "We want an inexpensive tail light made from 5 mm red LEDs. We want 160,000 mcd in stop mode, and 15% of that in tail light mode. It has to run over the normal automotive temperature range, −40 to +65 °C. We don't want it to dim even when battery voltage is low, the battery can be as low as 7 V during cranking. Power comes directly from the battery, so it needs to be OK with a 60 V load dump. We want lifetime to be at least 10,000 h. It needs to be a single string of LEDs; we can't have some LED strings on and others off."

This is all pretty specific. A rating in millicandela, however, doesn't tell us how bright it has to be without specifying the solid angle over which the light is distributed. They don't really understand the question. The answer appears to be that other companies are making tail lights this way, so whatever those other companies get is what our customer wants too.

## INITIAL ANALYSIS

We start by looking at 5 mm red LEDs in surface mount, limiting the search to those with ratings of at least 8000 mcd. That will require about 20 devices. Less brightness would require more LEDs, and that would increase the cost of assembling the board. (It costs about 1 ¢ per component to stuff.) Typing in "LEDs" on Digikey's website, we select "LED Indication – Discrete," as that is where cheap 5 mm devices are. We choose "red," and "In Stock," leaving 1741 choices. They range in intensity from 0.01 mcd up through 65,000 mcd. We choose those from 8000 mcd up, leaving 28 choices.

The cheapest one is from Cree, the C503B-RAN-CY0B0AA1, which runs $0.115 each at a thousand pieces. It shows a flux of 14,680 mcd at a test current of 20 mA. So we would probably end up using 11 of them, which will be $1.27, which is inexpensive, and probably 1/3 of that price in volume. This seems low enough that we

*Practical Lighting Design with LEDs*, Second Edition. Ron Lenk and Carol Lenk.
© 2017 by The Institute of Electrical and Electronics Engineers, Inc. Published 2017 by John Wiley & Sons, Inc.

won't try driving the LEDs' current harder than nominal to try and get the last millicandela out. For that matter, we may need to have a certain arrangement of the LEDs to make an attractive pattern. For example, we might want to end up with 12 LEDs instead of 11, so we can make a $3 \times 4$ array. We'll see.

These devices are 2.1 V forward voltage, so they're 42 mW each at test current, for a total power of 462 mW. That means an FR4 board will be fine. Eleven of these LEDs in series will be 23.1 V. Since the input voltage ranges from 7 to 60 V (load dump, see below), and the string voltage is between those two, the converter has to be a buck-boost. We'll use the HV9910 circuit discussed in Chapter 7. This has the considerable advantage that it can take 60 V directly on its input pin, so no additional circuitry is required to handle load dump. On the other hand, the HV9910 has a minimum operating voltage of 8.0 V (for the HV9910B; the HV9910C has a minimum of 15 V, which wouldn't be suitable at all), and there will also be some drop (probably a couple hundred millivolts) in the or'ring diode. (Stop power and tail light power come in on separate lines.) We will ask the customer if the cranking voltage requirement can be relaxed about 1 V.

This circuit will provide a constant current to the LEDs, but how much variation will there be in their light output over the temperature range? Red is notorious for changing a lot with temperature. The datasheet doesn't say. But we find a datasheet for another red made of the same semiconductor material, and we use that. It shows a curve indicating that the light at 65 °C is 75% as bright as at 25 °C. The curve only goes down to −30 °C, at which it is about 1.2× brighter. Extrapolating, we can guess that at −40 °C it is 1.3× brighter. Brighter is of course OK, the problem is when it's dimmer. So we could either add 1/75% = 1.33× more LEDs, or we could sense the temperature and compensate for the decreased light output when it's hot by increasing the drive current.

*Choice 1*: The extra 33% LEDs would be four devices. At $0.115 each that would be an extra $0.46. Since this is a 1000 piece price, this seems OK. What about board space? The devices are 5.8 mm in diameter, so we'll estimate total board space by giving them 6 mm × 6 mm = 36 mm$^2$ each. With 15 devices that will be a total area of 0.84 in.$^2$. That's OK; the board will be substantially larger than that anyway. The forward voltage will be increased to 31.5 V; we're still looking at buck-boost. The power is increased to 630 mW, which should still be OK thermally.

What about reliability of 15 LEDs? If they're all in series, and even one fails, then the tail light is out. Still, what's the alternative? If we have three strings of five LEDs each and one device goes out, the remaining two strings will produce only 2/3 the specified light. This doesn't meet regulations or customer specifications. Even worse, it may be detrimental if the driver is unaware that his tail light is dimmer than it's supposed to be. We'll just have to keep the LEDs cool enough that their MTTF can support the 10,000 h requirement.

*Choice 2*: The other choice would be to add in a temperature compensation circuit. As the temperature goes up, we would increase the current to the LEDs so that they get brighter, compensating for their drop in efficacy. We dislike thermistors for this application because of their nonlinearity. While you can add resistors around them to linearize it, and the accuracy isn't that important for this application, the real downfall of this method is consistent sourcing of the thermistors. As soon as the manufacturer substitutes in a "direct replacement" part, the characteristics are completely different.

Better is to use an IC. A quick glance shows some very inexpensive parts, for example \$0.19 at 1000 pieces for a Microchip MCP9700. So this will certainly be cheaper than four additional LEDs. On the other hand, it's going to take some circuitry around it. For example, the IC wants to run on 5 V and draws a maximum of 15 µA. We could zener it down from the battery. Zeners are available that will regulate with only 50 µA of current, so let's count 100 µA total. With a 7 V input during cranking, the resistor we need will be $(7 V - 5 V)/100 µA = 20 kΩ$. But now during load dump, this resistor has to dissipate $(60 V - 5 V)^2/20 kΩ = 150 mW$, so we need a 1/4 W resistor. At 14 V nominal battery voltage, we are throwing away $14 V^2/20 kΩ = 10 mW$, so it doesn't hit average efficiency much.

We also have to adapt the signal to the IC controller requirements. The MCP9700 generates 500 mV at 0 °C and this increases by 10 mV/°C. Its range for our application is then 100 mV at −40 °C to 1.15 V at +65 °C. Now in a typical LED driver, the LED current will be set by a sense resistor. A typical feedback voltage to an IC might be 250 mV, which is what it is for the 9910. Since we are putting 20 mA into the LEDs at 25 °C, the sense resistor should be $250 mV/20 mA = 12.5 Ω$. Now at +65 °C, the MCP9700 generates 1.15 V, and we need to adjust the current up to about $20 mA \times 1.33 = 26.7 mA$ using that 1.15 V signal.

To adjust the current up, we have to reduce the 12.5 Ω to a lower effective value. We can't just add the 1.15 V in to the sense node with a resistor somehow. Since 1.15 V is *higher* than the 250 mV feedback, so we would be *adding in* current. And adding current in *reduces* the LED current. We need instead to *increase* it. Looks like instead we need to invert the voltage, so that at higher temperatures it goes lower. We could build an opamp circuit, centered on 750 mV (the MCP9700's output at 25 °C), and have the difference add in to the sense node. That would also prevent the LEDs from getting brighter when it's cold. But to get the 750 mV will require a reference voltage. We could possibly divide down the 5 V supply we generated to run the MCP9700. . . . This is starting to get complicated, although it can probably be worked out. And we've just put another \$0.10 or more in the opamp.

One more consideration is that the MCP9700 takes 800 µs to turn on. Before it does, the circuit we're designing will see 0 V in its circuit, which it will interpret as −50 °C. Since this corresponds to very high efficacy for the LED, the driver will provide low current during this time. So for the first 1 ms, the LEDs will be a little dim. This is OK; we were just checking that it didn't flash at turn on.

Looking at the datasheet, we also see a curve "Output versus Settling Time to step $V_{DD}$." Apparently, when the supply voltage changes, the output takes some time to come back to its proper voltage. Output response to a 5 V step in the battery voltage looks to be about 700 ms. And in fact, it surges up to its power supply rail of 5 V. This seems potentially more serious: 5 V would be a temperature of 500 °C! And unless we have some kind of limit built in, the LEDs are going to see such a huge surge of compensatory current from the controller that they will blow up. Such a thing could actually occur. While we are using a zener to regulate the voltage to the device, a dip in the ground voltage, say due to a surge of current to a motor in the car, could cause the voltage actually seen by the IC to surge up. Placing the zener right next to the IC and adding an RC filter would mostly solve this, but it would take only one such incident to blow up the design.

This too can probably be fixed by adding a soft-start circuit to the controller response, but if it takes 700 ms to turn on to full brightness, then you're actually slower than an incandescent bulb! One of the advantages to LED tail lights is that they give the car behind you additional time to react. So something more complex is needed.

In the end, we decide that the design issues with temperature compensation are complex enough that we will just do the design with 15 LEDs. After the issues and costs are explained, the customer agrees to this, since that's what they wanted anyway.

## SPECIFICATION

With this information in hand, we draw up a specification for the customer, shown in Table 12.1. Let's go through it a line at a time. For both normal tail operation and for stop brightness, we have two lines of specifications. By the customer specification, we need to have a minimum stop brightness of 160,000 mcd (and 15% of this = 24,000 mcd for normal operation), so this has to be over both the full input voltage range and the full temperature range. Since it's in the minimum column, we need to make sure that it doesn't have to be tested. (Voltage range would be possible on a production line, but testing over temperature wouldn't be.) So we add in the words "by design."

On the other hand, we need to test *something* on the production line. So we select nominal temperature, and nominal operating voltage, which is about 14 V when

**TABLE 12.1 Specifications for LED Automobile Tail Light**

| Parameter | Min. | Typ. | Max. | Units |
|---|---|---|---|---|
| Stop brightness ($V_{in} = 14$ V, 25 °C) | | 195,000 | | mcd |
| Stop brightness ($V_{in}$ = full range, full temperature range, by design) | 160,000 | | | mcd |
| Tail brightness ($V_{in} = 14$ V, 25 °C) | | 29,300 | | mcd |
| Tail brightness ($V_{in}$ = full range, full temperature range, by design) | 24,000 | | | mcd |
| Emission angle, ½ brightness | | 15 | | ° |
| Dominant wavelength | | 624 | | nm |
| LED power | | 630 | | mW |
| Input power ($V_{in} = 14$ V) | | 740 | | mW |
| Ambient temperature (by design) | −40 | 25 | 65 | °C |
| LED temperature ($T_{amb} = 40$ °C) | | | 60 | °C |
| Input voltage | 8.5 | 14 | 60 | V |
| Board width | | 5 | | in. |
| Board length | | 2 | | in. |
| MTTF | | 10,000 | | h |
| LED cost (1 K pcs) | | | $1.75 | US$ |
| BOM cost (1 K pcs, incl. LEDs) | | | $4.25 | US$ |

the alternator in the car is running. This is then a typical number; we will add in the minimum and maximum values for testing once the worst-case analysis is completed.

The emission angle is just something we get from the LEDs' datasheet. It doesn't matter to us electrically, and it's certainly not something that's going to be tested in production, but the designer of the optics will need to know this. Dominant wavelength is in the same category, it's not something that will be tested, but it indicates to the customer that we're really using red, not, say amber.

LED power we estimated above as 42 mW × 15 LEDs = 630 mW. This is useful to us for the layout, to ensure thermal performance. If we assume 85% efficiency, input power will be 630 mW/85% = 740 mW, and this is something that can be tested on the production line. To make it concrete, we specify the voltage at which this is to be tested. And again, we'll add in limits after we do worst-case analysis.

Ambient temperature is the normal range for this automotive application. But again, since we can't test it, it's noted that it's by design. For LED temperature, what we're interested in is the MTTF. Naturally, the auto isn't at maximum temperature its whole life. That only happens when it's sitting in the sun in Florida in the summer! A conservative guess is that average ambient temperature over its life is 40 °C. We've taken a guess that 630 mW won't raise the temperature of the LEDs more than 20 °C. Since the part is rated at 100 °C, we hope that when we estimate the lifetime, 60 °C is enough to meet the specified MTTF.

We've documented the desired size of the tail light, as well as the desired MTTF. The LED cost is for thousand piece pricing. We've added in a guess of $2.50 for the price of the ballast components. The main driver of this is that the HV9910B is ~$0.80. (The HV9910C is only about $0.30, but remember that it doesn't work with the cranking voltage.)

As already noted, the input voltage with our design can't get down to the 7 V the customer originally wanted, since the IC won't run below 8 V. We've left ourselves a little margin by specifying that it will work down to 8.5 V. The customer accepts the 8.5 V minimum input voltage. We're given the go-ahead to design this system.

## LOAD DUMP

We've several times already mentioned load dump. What is it? Basically, it's a surge of voltage in an automobile's electrical system. What happens is this. The car battery isn't large enough to supply continuous power to all the electrical equipment in the car, such as the radio and, of course, the lights. So in normal operation, the car battery is kept charged by the alternator. The alternator is basically a generator. It has a magnet that is being spun by the engine, and that's inside a bunch of turns of wire. By Maxwell's equations, this generates an electric field, which in turn produces current through the wire. The current is fed into the battery, keeping it charged. And since it's attached to the battery, the voltage stays (approximately) constant.

Now of course "a bunch of turns of wires" is known as an inductor. And there's current flowing through this inductor. So what happens if the load is suddenly disconnected? It's exactly the same thing that happens in a boost converter: The output voltage goes up. And it goes up until there's a path for the current to flow through again, either because something like a resistor dissipates the energy, or until something breaks.

In automotive applications, this is what is called a load dump. If for some reason the battery is disconnected while the alternator is running, the output voltage surges up. While it's not exactly common, it does happen sometimes. A common cause is that you or a mechanic are working on your car with the engine running and accidentally remove the battery cable. The voltage flies up.

Now this happens often enough that car manufacturers typically add some surge protection. Nominal battery voltage is 12 V, and nominal alternator voltage is 14 V, so you might suppose the protection works at a reasonably low voltage. Unfortunately, there's a lot of energy in the alternator's winding, and we know that clamping devices have fairly soft $I$–$V$ curves. So the reality is that the load dump voltage can be as high as 60 V. This is of course better than the 120 V you might get *without* the surge protection, but it still means that the electronics must be able to survive the 60 V.

There are two ways of surviving load dump. The common way for most automotive electronics is to further clamp the voltage coming out of the surge protection. An easy way to do this is to put a resistor in series with a lower voltage clamping device such as a TVS diode. The resistor can be fairly large in resistance, if the current drain by the electronics isn't very large. And then the surge can be handled well by the clamping device, and the resistor just needs to be sized to take the power surge.

The other way of handling load dump is what we're doing with our circuitry. We have a SMPS that can run up to 60 V, and so it just rides through the load dump without dissipating any extra power. This latter method is more common in higher-power electronics, where the resistor required to adequately limit the current to the clamp would dissipate too much power during normal operation. Of course, our tail light is low power for a car, but we need a SMPS anyway to run the LEDs. Thus we get the benefit of surviving load dump without extra protection circuitry being needed.

## POWER CONVERSION

The circuit is shown in Figure 12.1, adapted from the circuit in Chapter 7. The first thing to notice is that it is *not* a buck-boost: it has one power transistor and one diode, not two of each. Nonetheless, it works both when the input voltage is lower than the LED voltage and when it is higher. This works because the bottom of the LED string is attached to the input voltage. The converter then produces a voltage higher than the input voltage at the top of the LED string. So it doesn't matter what the input voltage to the SMPS is, the LED voltage is always added on top of that voltage.

There are two inputs, "tail" and "stop." Remember that stop is supposed to be much brighter than tail. To accomplish these two different brightness settings, we've added in a circuit that detects when the stop power is applied. When stop is applied—but not when tail is applied—diode D3 conducts. This turns on small-signal transistor Q1. In that case, the current sense resistor R3 is put in parallel with R5, generating the higher current required for stopping. When TAIL power is applied, resistor R2 keeps

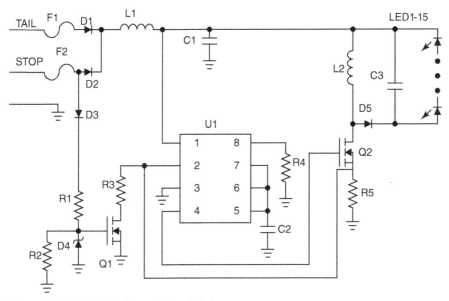

Figure 12.1   HV9910 schematic for tail light.

transistor Q1 off. R3 is thus disconnected from the circuit, and so R5 alone sets the current.

We've included fuses on both inputs because nothing should fail in such a way as to risk pulling large current from the battery. We've also included an input inductor L1 to form a basic EMI filter with C1. Since this is a SMPS, it's going to be generating EMI, and there are automotive specifications for the amount of noise we're allowed to generate. And we've included C3 to smooth out the current to the LEDs. The circuit produces pulses of current to the LED when transistor Q2 is off, and we want to get an LED current waveform without big peaks. Of course the switching frequency will be way too high to be visible, but large peak currents might possibly affect the lifetime of the LEDs.

Having said all that, it now occurs to us that although there are two power inputs, tail and stop, there's only one power output, namely ground. So if we put the fuse in the ground path, that covers both inputs, saving us one (expensive) component. It still works, because a short in the circuit means the power is coming from one of the two inputs, and in either case going to ground. And as for the EMI circuit, suppressing noise isn't nearly as good as not making it in the first place. The 9910 has a mode in which it will operate in constant off-time. As the input and/or output voltage changes, the on-time will also change, to compensate. This results in a varying frequency. That means the power in EMI will be spread out over more frequencies, lessening the amount of filtering needed. Now, for the automobile, the input and output voltage don't change very much on subsecond timescales. But still, it's free, and it might help a little bit, so we make that change as well.

We need to select values for F1 (now in the ground), C1 and C3 (C2 is always 100 nF), L1 and L2, R1 and R2, R3, R4 and R5, D1 and D2, D3, and D4 and D5. Let's start with R4 to set the frequency, as that determines the size of the inductor. For this design, we'll go for a switching frequency of approximately 50 kHz. This is a compromise between being far away from the 150 kHz edge of automotive EMI regulations and getting into very large component sizes. Remember of course that we've gone to constant off-time. A frequency of 50 kHz is a period of 20 μs. So if we choose an off-time of 10 μs, according to the datasheet we can use R4 = 250 kΩ.

Given the frequency, we can determine the inductor value. First, we will use volt-second balance across the inductor to determine the duty cycle. The LEDs' forward voltage is about $15 \times 2.1$ V $= 31.5$ V. Since the LEDs are in series with the inductor, this voltage is the voltage across the inductor while the transistor is off. When the transistor is on, the voltage across the inductor is the input voltage. The product of voltage and time has to be equal in these two conditions; otherwise the average current in the inductor would change. Thus,

$$31.5 \text{ V} \times t_{\text{on}} = V_{\text{input}} \times t_{\text{off}}.$$

By adding $V_{\text{input}} \times t_{\text{on}}$ to both sides, we can get a relation between the period $T$ and $t_{\text{on}}$:

$$31.5 \text{ V} \times t_{\text{on}} + V_{\text{input}} \times t_{\text{on}} = V_{\text{input}} \times t_{\text{off}} + V_{\text{input}} \times t_{\text{on}},$$

which is

$$(31.5 \text{ V} + V_{\text{input}}) \times t_{\text{on}} = V_{\text{input}} \times T$$

and since $t_{\text{on}}/T$ is by definition duty cycle,

$$DC = \frac{V_{\text{input}}}{31.5 \text{ V} + V_{\text{input}}}$$

When $V_{\text{input}} = 8.5$ V, DC = 21%, and when $V_{\text{input}} = 60$ V, DC = 66%. $T_{\text{on}}$, which is $D/f$, is then $t_{\text{on}} = 4.2$ μs at $V_{\text{input}} = 8.5$ V, and $t_{\text{on}} = 13.2$ μs at $V_{\text{input}} = 60$ V. We observe that at 8.5 V, the period is 10 μs + 4.2 μs = 14.2 μs = 70 kHz, which is still less than half of the 150 kHz where EMI is measured; and at 60 V, the frequency is 43 kHz, close to the 50 kHz we started with.

Since the inductance that the datasheet recommends is proportional to $t_{\text{on}}$, the biggest value occurs at $V_{\text{input}} = 60$ V. We end up with

$$L = \frac{(V_{\text{in}} - V_{\text{LED}}) \times t_{\text{on}}}{0.3 \times I_{\text{LED}}} = \frac{(60 \text{ V} - 31.5 \text{ V}) \times 13.2 \text{ μs}}{0.3 \times 20 \text{ mA}} = 63 \text{ mH}.$$

This relatively huge value is a direct consequence of our 50 kHz selection; but even if we had gone up a factor of 5, to 250 kHz, which is the practical limit of the IC, we would still have ended up with 5μmH. This is still huge because the current is low.

The 0.3 in the denominator of the inductor equation came from the datasheet assumption that the peak-to-peak ripple current in the inductor is 30% of average. Thus, peak current is going to be 15% higher than average, which is to say 23 mA. Looking for inductors between 47 and 100 mH, we see that all of our choices have at least 33 mA saturation current rating. We choose the cheapest one, a 47 mH, 30 mA part from Epcos, manufacturer part number B82144A2476. The rated resistance of the part is a surprisingly high 230 Ω, so we should check the power dissipation. At 20 mA this gives power dissipation of 92 mW, which seems like a reasonable number. The part is a cylinder akin to old-fashioned through-hole resistors, with a diameter of 0.205 in. and a length of 0.472 in. It has a 10% drop in inductance at 33 mA, so it should be fine at 23 mA. Since all of our parts are going to be automotive spec, we are going to use the straight distributor 1000 piece price, triple the normal volume pricing we assume. This inductor is $0.303.

Next we turn to the transistor Q2. Maximum input voltage is 60 V, and the LEDs add another 30 V on top of that. It's easy to pick a 200 V MOSFET, so we go for low price and small package. The ZVN3320 from Zetex lists at $0.166 in 1000 pieces, and is in a SOT-23, nice and small. It's rated at 25 $\Omega$, 60 mA continuous. With 20 mA of current and a 66% duty cycle, the on-state losses will be only $(20\,\text{mA})^2 \times 25\,\Omega \times 66\% = 7\,\text{mW}$.

We can also choose the diode D5. Since both the current and the frequency are low, and the output voltage across the LEDs is high, diode power loss will be small, so we don't need a Schottky. We can select D5 to have the same rating as Q2, 200 V. We look for a 1 A part since those are common, and since they all pretty much have the same forward voltage, we pick an SMA package. We want something that isn't too fast, to avoid EMI problems. After all, the switching frequency is only 50 kHz. The RS1D seems suitable. Its power loss is only $650\,\text{mV} \times 20\,\text{mA} \times 79\% = 10\,\text{mW}$. Since the current is similar for D1 and D2, we pick the same part for these. The 200 V rating makes it good for protection against polarity reversal (in case the battery gets hooked up backward in the car). Since D3 doesn't carry any current, we choose a surface-mount 4148 for it, still giving enough voltage for polarity reversal protection.

We should also be ready now to pick the current sense resistors. We want 20 mA, with 6 mA peak-to-peak ripple, and so the resistor value should be 250 mV/ $(20\,\text{mA} + 0.5 \times 6\,\text{mA}) = 10.87\,\Omega$ when both resistors are in parallel. In tail light mode, the current should be $[20\,\text{mA} + (0.5 \times 6\,\text{mA})] \times 15\% = 3.45\,\text{mA}$. R5 alone should thus be 250 mV/3.45 mA = 72.5 $\Omega$. The closest standard value rounding down is R5 = 71.5 $\Omega$. (We round down to ensure that the LEDs will be *brighter*.) To get down to 10.87 $\Omega$ in stop mode, we need to parallel $(71.5\,\Omega \times 10.87\,\Omega)/(71.5\,\Omega - 10.87\,\Omega) = 12.82\,\Omega$. Now, this resistance is the series combination of Q1 and R3. We want a Q1 resistance to be small compared to this, so that it doesn't affect the current much. Looking for something in the 20–30 V $V_{DS}$ range, we find the DMG3420U-7, which is rated 25 m$\Omega$ with 4.5 V on its gate. So for R3 we can select the rounded down standard value 12.7 $\Omega$.

Since Q1 needs only 5 V on its gate to be fully enhanced, we choose D4 to be MMSZ5231. This takes only 50 μA to regulate at 5.1 V. Let's add in another 10 μA for R2; its purpose is only to ensure that Q1 is off if stop isn't powered. At 5 V, that corresponds to R2 = 5 V/10 μA = 500 kΩ. The standard value is 499 kΩ. If stop is at 8.5 V, R1 can be as large as (8.5 V − 5.1 V)/(50 μA + 10 μA) = 57 kΩ. If we choose 49 kΩ, then at 60 V it will dissipate $(60\,\text{V} - 5.1\,\text{V})^2/49\,\text{k}\Omega = 62\,\text{mW}$, so a 0805 will work fine.

We still need to pick C3. This capacitor has to provide energy to the LEDs during the time the MOSFET Q2 is on, since during that time the inductor current is shunted to ground. Worst case is when the MOSFET is on 66% of

$20\,\mu s = 13.2\,\mu s$. During that time, the capacitor is providing $20\,mA$. The charge delivered is thus $Q = I \times t = 13.2\,\mu s \times 20\,mA = 264\,nC$. We want the LED current to remain approximately constant. So let's say the LED voltage shouldn't drop more than $20\,mV$ per LED, or a total of $300\,mV$ for 15 LEDs. Since $Q = C \times V$, we have $C = Q/V = 264\,nC/300\,mV = 0.88\,\mu F$. We want $0.88\,\mu F$ at $31.5\,V$, and round up to $1\,\mu F$ at $50\,V$ (we don't choose $35\,V$ as being too little margin). We find parts as small as 0603. Selecting only parts that are at least $125\,°C$, the least expensive part is in 0805.

This leaves the input filter and the fuse for design. The purpose of C1 is to provide high-frequency bypass for the converter. It should probably be ceramic, even at $50\,kHz$. Since it sees up to $60\,V$, we'll want either an 80 or $100\,V$ cap. And now we can calculate its capacitance in the same way we just did for C3. We're going to pull $20\,mA$ from it for $20\,\mu s$ (worst-case), which is $400\,nC$. The lowest voltage input is $8.5\,V$, which we don't want to drop below, say, $8.4\,V$, $100\,mV$ drop. So capacitance should be at least $400\,nC/100\,mV = 4\,\mu F$. Our choices on Digikey turn out to be either $3.3\,\mu F$ or $4.7\,\mu F$, and choosing the latter, the cheapest turn out to be in a 1210. As a check, our rule-of-thumb for capacitors says that $4.7\,\mu F$ will work up to $1\,MHz$-$\mu F/4.7\,\mu F = 200\,kHz$, so that's an OK value.

Can we use the same inductor for the input filter as for the power conversion? No, the current is much higher. We're pulling $740\,mW$, and the minimum input voltage is $8.5\,V$, so at that point the converter is pulling $83\,mA$. The power conversion inductor pulls a lot less current because it's at a much higher voltage ($31.5\,V$ across the LEDs). So let's quickly go through the selection process. We want a rated current between $83\,mA$ and say $120\,mA$. Then we'll look for inductance in the range of $10$–$100\,mH$. And finally, we choose the smallest nine sizes to take a look at. The largest inductance value in this set is $33\,mH$, a Bourns inductor part number RLB9012-323. (It seems suspiciously like this part number should indicate an inductance of $32\,mH$, but that's what the datasheet says.) And its cost is actually less than the $47\,mH$ we're using for the power conversion, so we use this. We get a roll-off frequency of $1/(2\pi \sqrt{LC}) = 1/[2\pi \sqrt{(33\,mH \times 4.7\,\mu F)}] = 404\,Hz$. This is a factor of $50\,kHz/404\,Hz = 124 = 42\,dB$ below the switching frequency, and so is a good choice for the filter inductor as well.

Finally, for the fuse we find a problem. Our distributor carries automotive fuses, but the lowest current is 1A, which seems very high for a circuit that is only supposed to be pulling $20\,mA$. That's 50× higher than rating! A quick call to the customer reveals that this component is already part of the assembly driving the tail light, so we leave it off the BOM, and will remove it from the schematic.

Our second-pass schematic and BOM are shown in Figure 12.2 and Table 12.2. The PCB price is $0.21, which is only $0.021/in.$^2$, because our customer is having the board made by its volume manufacturer in China. They must have some volume pricing deal with them. We've remarkably ended up almost exactly on budget, both on the LEDs (despite increasing their quantity) and on the rest of the BOM. The most expensive items, in order, are the LEDs, the controller, the inductors, and then the input cap. We recommend to the customer talking to the multiple vendors for the controller to get the price down even further, and checking with their Chinese

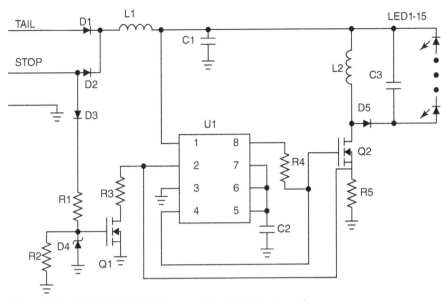

Figure 12.2    Final HV9910 LED automobile tail light schematic.

**TABLE 12.2    LED Automobile Tail Light BOM**

| Ref des | Description | Part # | Mfr. | 1 K Price |
|---------|-------------|--------|------|-----------|
| C1 | 4.7 µF, 80 V, 1210 | Any | Any | $0.284 |
| C2 | 100 nF, 10 V, 0805 | Any | Any | $0.043 |
| C3 | 1 µF, 50 V, 0805 | Any | Any | $0.017 |
| D1 | 200 V, 1 A Ultrafast, SMA | RS1D | Any | $0.076 |
| D2 | 200 V, 1 A Ultrafast, SMA | RS1D | Any | $0.076 |
| D3 | MMBD4148, SOT23 | Any | Any | $0.039 |
| D4 | 5.1 V Zener, SOT-23 | MMSZ5231 | Any | $0.039 |
| D5 | 200 V, 1 A Ultrafast, SMA | RS1D | Any | $0.076 |
| L1 | 33 mH, 90 mA, 92 Ω | RLB9012-323KL | Bourns | $0.245 |
| L2 | 47 mH, 30 mA, 230 Ω | B82144A2476 | Epcos | $0.303 |
| Q1 | 20 V, 29 mΩ, SOT23 | DMG3420U-7 | Diodes Inc. | $0.081 |
| Q2 | 200 V, 25 Ω, SOT23 | ZVN3320 | Zetex | $0.166 |
| R1 | 49 kΩ, 0805 | Any | Any | $0.004 |
| R2 | 499 kΩ, 0805 | Any | Any | $0.004 |
| R3 | 12.7 Ω, 0805 | Any | Any | $0.004 |
| R4 | 249 kΩ, 0805 | Any | Any | $0.004 |
| R5 | 71.5 Ω, 0805 | Any | Any | $0.004 |
| U1 | Controller | HV9910B | Supertex | $0.690 |
| PCB | | | | $0.210 |
| Sub-total | | | | $2.365 |
| LED1-15 | 624 nm. 14,680 mcd LED | C503B-RAN | Cree | $1.725 |
| Total | | | | $4.09 |

manufacturer about cheaper replacement parts for the caps and inductors. They accept the design as is.

## THERMAL MODEL

It doesn't seem like very much power dissipation for such a big board, but on the other hand we've designed it so that the LED temperature rise should be small. We'll again use Chapter 6 to form an estimate. We have 740 mW of dissipation including the driver. The surface area is 5 in. × 2 in. = 10 in.$^2$. Power density is thus 74 mW/in$^2$. Looking at the curve in Chapter 6 for a 40 °C ambient, the temperature rise estimate is 10 °C, which takes us to 50 °C, 10 °C lower than the maximum temperature we had allocated.

But hold on! That's the board temperature; what we want to know is the LED junction temperature. Now, the datasheet doesn't actually list junction-to-case thermal resistance. But looking carefully, we find a thermal curve showing maximum forward current versus ambient temperature. It states (in parentheses!) that maximum junction temperature is 110 °C. The curve starts to roll-off from 50 mA allowed current at 60 °C. And another curve on the datasheet shows that at 50 mA we have a forward voltage of 2.35 V. So the thermal resistance is (110 °C − 60 °C)/(50 mA × 2.35 V) = 430 °C/W. We recall that each LED is dissipating 42 mW. So the junction rises 430 °C/W × 42 mW = 18 °C, and the junction temperature is thus 68 °C. This is 8 °C hotter than we targeted.

We'll have to add some heatsinking to the design. The effective thermal resistance of the design right now is 10 °C/74 mW = 135 °C/W. We need the temperature rise to be one-fifth of what it is with just the air, so we need an additional thermal path of 27 °C/W. Fiddling around with some heatsink options, it looks like we're going to need a custom extrusion. This starts to sound worrisome, until it's pointed out that we forgot that we are already attached to a really big heatsink: the vehicle! All we need to do is to find an electrically insulating thermal interface, and bolt the board down through it.

We go the Bergquist website (www.bergquistcompany.com), and look for thermally conductive insulators. The one called "economical," the Sil-Pad® 1100ST, says it is fiberglass-reinforced and is suitable for automotive electronics. The thermal conductivity at rated contact pressure is 1.1 W/(m K). To find the thermal conductivity, we multiply by the area and divide by the thickness of the material, 1.1 W/(m K) × (0.0065 m$^2$)/0.00031 m = 23 W/K. The thermal resistance is the inverse of this, 0.043 K/W, so low that we are way overdoing it. The LEDs will just be sitting at ambient temperature. Similar materials at a distributor are priced at $48.66 for a 10 in. × 12 in. pad. This is big enough for 12 full boards, and using the divide-by-three rule we still end up with a cost adder of $1.35 per board. This is half as much as the whole ballast! So in the end, we decide to cover only the ballast part of the board, since this is the only part with live connections exposed on the backside anyway. This is about 1½ in. × 1½ in. = 2¼ in.$^2$ = 1/5 of the board with this material, giving us five times the thermal resistance (still less than 1 W/K) and 1/5th the cost ($0.27), which seems like an acceptable trade-off. The part of the board without the Sil-Pad can be insulated with fiberglass, or any other inexpensive electrical insulator.

# WCA

The key specification the customer is interested in is the brightness of the light. We've designed the system to meet those specifications, but a lot of engineering approximations went in to get there. For example, resistor values are nominals; we didn't consider what happens if they come in at one or the other end of their tolerance specification. Worse, we didn't consider what happens if all the resistors come in at the ends of their tolerance. While this may seem like it would require a malicious conspiracy on the part of manufacturers, if you build enough of something eventually something like this will happen.

Now there are two ways to approach this question. (Well, actually three: You could ignore it and hope that it's OK. But that's a terrible approach to engineering, and will catch up with you someday.) One way would be to *estimate* how many units are going to have problems. In this approach, known as Monte Carlo, you use a computer to simulate hundreds (or thousands) of different sets of component values in your design. Then you graph the results. The percentage of simulations that fail the specification can be assumed to be equal to the percentage of real units that would fail.

This would be fine, except for a hidden, implicit assumption. Suppose you have designed in a $10\,k\Omega$ resistor, and its tolerance is 5%. That means the parts can be anywhere between 9500 and $10,500\,\Omega$. It might seem natural to assume that the value $9800\,\Omega$ is equally likely as $10,200\,\Omega$. *But that isn't necessarily true*! In fact, you've assumed a flat probability distribution function. That means you've assumed that every value in the range is equally likely as every other value. Now if the value of the resistance is subject to many different random effects, then the Central Limit Theorem may apply, and the distribution of values may in fact be evenly spread over the tolerance range.

But that's not the way components are *really* built. The reality is that when you build a device, it has a certain mean value, and then a distribution of values around that mean. The mean value can be anywhere in the tolerance range. The distribution around that range is usually very tight. So if you buy a reel of three thousand $10\,k\Omega$ resistors, the mean value may be, let's say, $9800\,\Omega$. And the distribution will be very tight—almost all of those resistors will be between 9700 and $9900\,\Omega$. If it turns out that values between 9700 and $9900\,\Omega$ cause a big problem for your circuit, then it's not true that $(9900\,\Omega - 9700\,\Omega)/(10,500\,\Omega - 9500\,\Omega)$ 20% of your board will have the problem. *Almost all* of the boards will have the problem!

So for this reason we dislike using probabilistic assessment of robustness. You just don't know what the probability of getting a particular value is. The second method of assessing whether your design meets specifications gets around this problem by evaluating the worst possible case. For that reason, it's known as worst-case analysis, or WCA. You evaluate each component's worst possible value, and assume that they all maliciously happen at the same time on a single board. So in the case of the brightness of our light, we're going to evaluate every component's effect on the brightness. We'll look at minimum and maximum values, including temperature variation, and select whichever value is worst. And then we'll assume that all of these worst values occur at the same time. (Except of course that the temperature can't be both maximum and minimum at the same time!) And that will tell us what the minimum brightness is, and whether it meets specifications. If it doesn't, it usually turns out that just one or two components are the reason, and we can easily go back in and adjust those. Then we can be confident that 100% of our build will meet specification (setting aside manufacturing problems).

In this sample analysis, we're going to be looking just at the minimum brightness produced by our light. In a more thorough analysis, we would also want to know things such as the maximum power dissipated in components, maximum voltage stresses, and so on. We'd also want to calculate maximum temperature of components (in worst-case ambient temperature), to see if we're exceeding any ratings. Note that this maximum temperature is *not* the one we would use for MTTF. That temperature should be the average operating temperature.

So to start with, we need to identify which components affect the brightness. Obviously, the LEDs have a key role. Their brightness depends not only on their intrinsic construction but also on the current passing through them. That current is determined by the current sense resistors, and by the feedback threshold voltage of the IC. There are also, presumably smaller, effects on the current. For example, once the voltage across the sense resistor reaches the feedback voltage threshold, it takes some finite time for the IC to respond, and then for the MOSFET to turn off. The MOSFET turn-off time is probably so fast that it doesn't matter, but we'll evaluate the gate charge just in case.

Having identified the parameters of interest, we need to construct a table, as shown in Figure 12.3, showing worst-case values of each parameter. Our first column lists the reference designator, and the second column shows which parameter of the component we're interested in. Then we list the nominal value of the parameter, its units, and its tolerance. Using those two, we calculate minimum and maximum values at room temperature in the next two columns. For example, in the first line we consider the case of the value of the resistance of R5, the current sense resistor. This is of course of direct importance in setting the brightness. We have $71.5\,\Omega - 1\% = 70.785\,\Omega$ for the minimum value, and $71.5\,\Omega + 1\% = 72.215\,\Omega$ for maximum.

In the next column, we list the temperature coefficient of the parameter from the component's datasheet. R5's resistance's temperature coefficient is $\pm100\,\text{ppm}/°C$. Since such numbers are always relative to the value at 25 °C, we need to know the maximum deviation of the components from that temperature. In many cases, the

|   | A | B | C | D | E | F | G | H | I | J |
|---|---|---|---|---|---|---|---|---|---|---|
| 1 | Ref. Des. | Parameter | Nominal | Units | Tolerance | + | − | Tempco | + | − |
| 2 | R5 | Resistance | 71.5 | Ω | 1% | 72.215 | 70.785 | ±100ppm/C | 72.68 | 70.32 |
| 3 | R3 | Resistance | 12.7 | Ω | 1% | 12.827 | 12.573 | ±100ppm/C | 12.91 | 12.49 |
| 4 | Q1 | Resistance | 25 | mΩ |  | 35 |  | 1.25 | 43.75 |  |
| 5 | Q1 | IGSS | 1 | µA |  |  |  | 400% | 4 |  |
| 6 | U1 | VCS, TH | 250 | mV | 10% | 275 | 225 | none | 275 | 225 |
| 7 | LED1-15 | IV | 14,680 | mcd |  | 23,500 | 12,000 | 1.3, 0.65x | 30,550 | 7,800 |

Figure 12.3 WCA table.

individual components will be self-heating, so that it is necessary first to estimate their power dissipation, and then their temperature in order to determine their actual operating temperature. Of course, this process is iterative. In the case at hand, we don't expect any components to rise above the LED temperature. Since we have calculated them to have a 20 °C rise (60 °C with a 40 °C ambient), and since maximum ambient temperature is supposed to be 65 °C, we will suppose that the hottest temperature is 65 °C + 20 °C = 85 °C. At the other end, the coldest temperature will be at the minimum ambient of −40 °C, at turn on, before any of the components have time to self-heat. Maximum deviations are thus 85 °C − 25 °C = +60 °C and −40 °C − 25 °C = −65 °C.

The maximum resistance of R5 will occur when it is already at the top of its tolerance—72.215 Ω—plus having a negative temperature coefficient of −100 ppm/C at the maximum temperature deviation of −65 °C. We thus have 72.215 Ω × [1 + (−100 ppm/°C) × (−65 °C)] = 72.68 Ω. Similarly, the minimum resistance occurs when the value of R5 is minimum and we have a positive temperature coefficient and the maximum temperature deviation, giving 70.32 Ω. We can do exactly the same calculation now for R3, the paralleled current sense resistor, as shown in the next row. It has the same tolerance and temperature coefficient as R5, so it's simple to cut and paste the formulae. R3 ranges from 12.49 to 12.91 Ω.

R3 is in series with Q1, which is the switch controlling R3. Could the resistance of Q1 matter? Of course, the resistance can't be negative, so it can only increase the maximum resistance of the current sense with R3, not decrease it. So we only calculate the maximum resistance of Q1. The datasheet for Q1 says that at 4.5 V gate-source, maximum resistance is 35 mΩ, and that at 10 V it's 29 mΩ. We're actually driving it at 5.1 V, so the 35 mΩ is the better choice for our calculation. Next, the resistance of course goes up with temperature. Our maximum temperature is 85 °C, at which temperature the datasheet curve shows that the resistance increases by 25%. Multiplying through, we find the maximum resistance of Q1 to be 44 mΩ. This is in the fourth decimal place for R3, and so won't be important.

Another consideration comes to mind. When stop isn't powered, Q1 is supposed to be off, so that R3's resistance to ground is infinite. But of course, transistors leak current even in the off state, and that current increases with temperature. Could this possibly affect the value of the resistance in parallel with R5? Looking at the datasheet for Q1, maximum leakage current at 25 °C is 1 μA. (It's actually probably a lot less, and this is just the limit of their testing equipment.) A curve in the datasheet shows that at 85 °C, the leakage increases about four times, so maximum leakage current will be 4 μA. Going back to the conditions at R5, there's supposed to be 20 mA of current through the LEDs and thus through R5. The leakage current of Q1 thus represents only 4 μA/20 mA = 0.02%, so we can ignore it.

Next, we need the feedback voltage on the IC, since that sets the voltage sensed by the current sense resistors. The datasheet shows that this voltage has a tolerance of ±10% over our whole temperature range. Minimum feedback voltage is thus 225 mV, right from the datasheet without calculations. We fill in the table to record the data.

The last thing we need is the brightness of the LEDs. First we have to find the flux bin information. It turns out, rather alarmingly, that the flux bins are 2× wide— from 12,000 to 23,500 mcd! Fortunately, nominal, 14,680 mcd, is close to the bottom end of the flux bin. We'll have to see how far off we are. We also need the temperature variation of the flux. From our original calculations, we already know that light at $-40\,°C$ is 1.3× brighter. But for the high temperature end, we estimated brightness at $65\,°C$, rather than $85\,°C$. Going back to our source, extrapolation suggests that the brightness is 65% of room temperature at $85\,°C$. Plugging the numbers in, brightness will be a minimum of 7800 mcd—we don't care about the maximum.

We now have everything we need to find the minimum brightness of our design. Minimum brightness occurs when the current through the LEDs is minimum. That occurs when the current sense resistors have their maximum resistance and the feedback voltage is minimum. So first, the maximum resistance in stop mode is [R5 in parallel with (R3 plus Q1)] = $[72.68\,Ω \times (12.91\,Ω + 0.04\,Ω)]/[72.68\,Ω + (12.91\,Ω + 0.04\,Ω)] = 10.99\,Ω$. That means the IC will turn off Q2 at a minimum current of $225\,mV/10.99\,Ω = 20.47\,mA$.

The LED is was calculated to have a minimum flux of 7800 mcd at 20 mA. At 20.47 mA, that will be 7983 mcd. We have 15 LEDs, giving a minimum brightness of 119,749 mcd. That's a lot less than the 160,000 we were expecting! What happened? It's really a combination of two effects. The temperature is higher than we originally estimated, which shaved 10% off our estimate, and then we didn't account for the flux binning at all, which took another 20% off. Altogether, we need 34% more flux!

What are our choices? We probably can't go back to the customer and tell him that some of the units won't be bright enough. We could increase the current to the LEDs by 34%, taking it up to 26.8 mA. Or we could increase the number of LEDs by 34%, taking it up to 20 LEDs. Which is easier/cheaper? Clearly adding LEDs involves extra cost. Five extra LEDs adds $0.57 to the BOM. That takes us $0.47 over our original estimate, which however is only a 10% bump.

Increasing the LED current would just be resistor changes, but we need to check the datasheets for both the LEDs and the inductor, to make sure they're OK with the extra current. The LED says it has an absolute maximum current of 50 mA, so that's OK. For 26.8 mA with 6 mApp ripple current gives 29.8 mA peak current in the inductor, which was rated at 30 mA, so that's also OK.

Since the extra current works, whereas the extra LEDs costs money, we choose the former. We rescale both R5 and R3 down by 119,749 mcd/160,000 mcd because lower resistance means higher current. We end up with R3 = $9.5\,Ω$ and R5 = $53.5\,Ω$, and the closest standard values being 9.53 and $53.6\,Ω$, respectively. We'll leave checking the brightness for the tail mode for the reader. So our WCA analysis was worthwhile—we caught and corrected what would otherwise be a big goof before even going to layout!

## TESTING

Since the customer has a critical parameter in the design—the brightness—it would be worth thinking about how to test that. On the production line, stuffed units are going to be rolling out after soldering, and we need to make sure that each unit is bright enough. Deciding which things to test and how to cost-effectively test them is sometimes done by a dedicated test engineer, but we'll take on that role as well.

The way to think about this problem is to consider what things could go wrong, causing the brightness to be wrong. We said "wrong" rather than just "too low" because we really need to also find out if the brightness is way too high. If the current through the LEDs is too high, other things might burn up. For example, the inductor doesn't have that much current rating above where it's being operated. If the current is too high, it will start to saturate. Of course, if the current is *way* too high, that will simply blow the fuse. But if it's just *somewhat* too high, it may cause overheating, in the inductor as well as in the transistor, the current-sense resistor, and the LEDs.

The most straightforward way to test the brightness of the LEDs would be to measure each one with a lux meter. We could construct a little test fixture so that the distance from the unit to the board is always the same. Since the lux is inversely proportional to the square of the distance, you wouldn't want any variations in distance causing incorrect readings. And then if we could find a lux meter with a digital bus output, maybe something like an Ethernet cable, we could hook it up to an automatic testing machine. The conveyor belt would advance each unit coming off the production line into the spot directly beneath the lux meter, make a measurement, and shunt the unit either into the packaging line, or into a box marked "rework" if the measurement result is too high or too low.

This would be a fine plan if we were planning to make millions of these. The problem is that all this machinery and construction takes a fair amount of time and quite a bit of money. And it doesn't give the rework people any clue as to what might be the problem. It would be better to find a hand-operated test procedure that could also give some indication of what's likely to be wrong.

The most common failure on LED boards is, believe it or not, for the LEDs to be put in backward. Fortunately, this is an easy thing to test for. If any LED in the string is backward, none of them will light up. So if we apply a small current to the string, say 5 mA, they should all light up. This can be done by hand with a lab power supply. We just need to add test points to the PCB so there's a place to touch with probes. The test is run without the input power, since we want to ensure that the LED load is attached before powering the power supply up. A boost converter doesn't fare well without a load.

The other thing affecting the brightness is of course the current through the LEDs. One common proxy for the current is to measure the input power to the unit. With a fixed input voltage (say nominal 14 V), you can measure the power input, and it should be always the same for every board manufactured, say within $\pm 5\%$ or better. This is a great test to see that nothing's shorted or open, and in many cases may be all that's needed. In this particular case, we decide we want to be a little more specific. We'll put a 1 Ω resistor in the LED string. When there's 20 mA going through the string, the voltage across the resistor should be 20 mV. And since there's a capacitor across the string, the voltage can be measured accurately just using a DMM. We add the resistor, plus one additional test point for probing. It was useful to do the test plan before the layout was finished, so that we could get the extra component and test points in without upsetting the schedule.

## PCB

Passing our schematic over to the layout engineer, we get back Figure 12.4. Remember that we changed the values of R3 and R5 during WCA. The three connectors have again been explicitly called out so that there will be pads for them in the layout. And similarly, the test points also are explicit, so that they will have pads as well.

The PCB for the entire assembly is shown in Figure 12.5. The topside is shown in red and the bottom in aqua. On the top, we see the LED layout. We wanted to make a symmetrical distribution of the light sources, and also come close to the edge of the board to make sure the light-emitting surface was as large as possible. Of course the number of LEDs, 15, isn't a prime, so we can lay out the LEDs as the prime factors of 15.

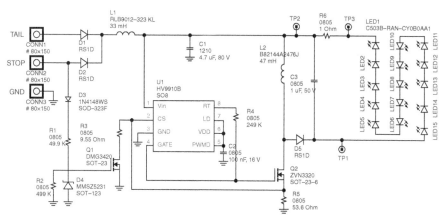

Figure 12.4   HV9910 LED automobile tail light schematic.

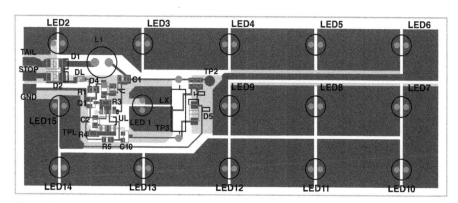

Figure 12.5   LED tail light. Whole board.

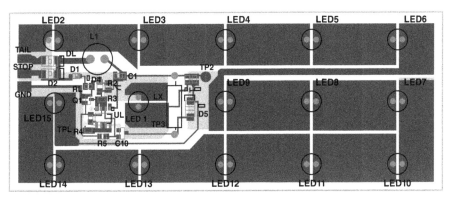

Figure 12.6   LED tail light. Top view. (See color plate section.)

What we've done here is made an array of $3 \times 5$ LEDs. This is the most symmetric arrangement possible. There is a pad for connection on the far left for the input powers and ground. The fuse isn't included, as decided earlier.

In Figures 12.6 and 12.7, we show the top and bottom sides, respectively. The traces connecting the LEDs are large. In fact, they are copper pours completely using all the space available. With only 20 mA current, they didn't need to be so large, and they didn't really need the copper for heat dissipation either. Instead, we filled up the space to help the mechanical rigidity of the board. For the same reason, the LED connections have pours on the backside as well. Of course, the top and bottom pads are connected by the leads of the LEDs.

Notice that with the connectors all at the far left, the layout of the LED pads requires their connection from the lower left-hand corner to the upper left-hand corner, without connecting to the other two corner pads. Going back and forth three times (three rows) would have ended up with one of the ends of the LED string on the wrong side. Instead, the layout engineer has added the trace between LEDs 6 and 7 to do the crossover, without having to run a trace the whole length of the board. This required some thought to get the array properly organized.

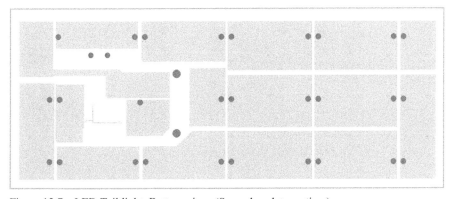

Figure 12.7   LED Tail light. Bottom view. (See color plate section.)

In the middle of the left side of the layout you can see the ballast part of the circuit. As promised, it is somewhat smaller than 2 in. × 1 in. It is all colocated, so the thermal pad doesn't have to be too large. The positive input connectors come in through the two diodes that are joined immediately at their cathodes. They then go directly to the input filter inductor L1. Although L1 is through-hole, we don't use it to get to the backside, as it wasn't necessary. The other end of L1 is power to the system, and goes by means of a trace on the topside to L2, C1, C3, and, most importantly, through R6 to LED1. The other end of L1 goes to the switching transistor Q2 and, more closely, the power rectifier D5. The cathode of D5 then goes over to the start of the LED string, the anode of LED15.

Ground is done much the same way. It comes in on the trace on the topside. This then connects right away to the source of Q1, which is the highest ground current in the system. It then picks up the ground of the IC and the bypass capacitor for the reference voltage of the IC, C2. Finally, it picks up R5, the current sense resistor for the switching; this is 15% of the current from Q1, and so can be somewhat further away. Observe that C1, the bypass power for the switcher, has its ground routed back to the power ground of the IC.

**TABLE 12.3 Final LED Automobile Tail Light BOM**

| Ref. des | Description | Part # | Mfr. | 1 K rice |
|---|---|---|---|---|
| C1 | 4.7 µF, 80 V, 1210 | Any | Any | $0.284 |
| C2 | 100 nF, 10 V, 0805 | Any | Any | $0.043 |
| C3 | 1 µF, 50 V, 0805 | Any | Any | $0.017 |
| D1 | 200 V, 1 A Ultrafast, SMA | RS1D | Any | $0.076 |
| D2 | 200 V, 1 A Ultrafast, SMA | RS1D | Any | $0.076 |
| D3 | MMBD4148, SOT23 | Any | Any | $0.039 |
| D4 | 5.1 V Zener, SOT-23 | MMSZ5231 | Any | $0.039 |
| D5 | 200 V, 1 A Ultrafast, SMA | RS1D | Any | $0.076 |
| L1 | 33 mH, 90 mA, 92 Ω | RLB9012-323KL | Bourns | $0.245 |
| L2 | 47 mH, 30 mA, 230 Ω | B82144A2476 | Epcos | $0.303 |
| Q1 | 20 V, 29 mΩ, SOT23 | DMG3420U-7 | Diodes Inc. | $0.081 |
| Q2 | 200 V, 25 Ω, SOT23 | ZVN3320 | Zetex | $0.166 |
| R1 | 49 kΩ, 0805 | Any | Any | $0.004 |
| R2 | 499 kΩ, 0805 | Any | Any | $0.004 |
| R3 | 9.53 Ω, 0805 | Any | Any | $0.004 |
| R4 | 249 kΩ, 0805 | Any | Any | $0.004 |
| R5 | 53.6 Ω, 0805 | Any | Any | $0.004 |
| U1 | Controller | HV9910B | Supertex | $0.690 |
| PCB | | | | $0.210 |
| TIM | | | | $0.810 |
| Subtotal | | | | $317.5 |
| LED1-15 | 624 nm, 14,680 mcd LED | C503B-RAN | Cree | $1.725 |
| Total | | | | $4.90 |

# FINAL DESIGN

We end up with this BOM, Table 12.3. The price is \$0.81 higher yet, \$0.65 over our original estimate. This is mostly because of the Sil-Pad (TIM stands for thermal interface material). The customer seems to expect this sort of price over-run. We make a note to ourselves to look up how much retail pricing is on the Internet. The volume pricing will only be \$0.27 higher anyway. And we know that cheaper LEDs are on their way.

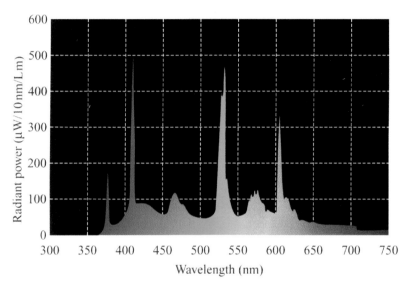

Figure 1.2    Fluorescent tube's spectral power distribution. (*Source:* http://www.gelighting
.com/na/business_lighting/education_resources/learn_about_light/pop_curves.htm?1.)

*Practical Lighting Design with LEDs*, Second Edition. Ron Lenk and Carol Lenk.
© 2017 by The Institute of Electrical and Electronics Engineers, Inc. Published 2017 by John Wiley & Sons, Inc.

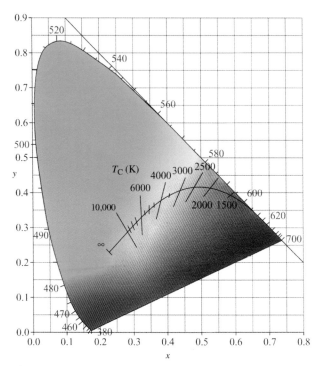

Figure 3.11    CIE 1931 $(x, y)$ chromaticity space, showing the Planck line and lines of constant CCT. (*Source:* http://en.wikipedia.org/wiki/Color_temperature under license http://creativecommons.org/licenses/by-sa/3.0/.)

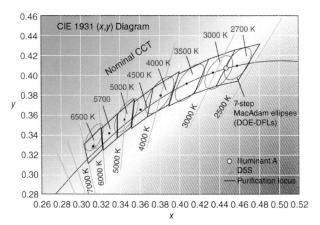

Figure 3.12    $(x, y)$ Chromaticity diagram showing CCT and seven-step MacAdam ellipses. (*Source:* http://www.photonics.com/Article.aspx?AID=34311.)

Figure 3.13 (a) Cool white fluorescent 4100 K, CRI 60; (b): Incandescent, 2800 K, CRI 100; (c): Reveal® incandescent 2800 K, CRI 78.

| Name | Appr. Munsell | Appearance under daylight | Swatch |
|------|---------------|---------------------------|--------|
| TCS01 | 7,5 R 6/4 | Light grayish red | |
| TCS02 | 5 Y 6/4 | Dark grayish yellow | |
| TCS03 | 5 GY 6/8 | Strong yellow green | |
| TCS04 | 2,5 G 6/6 | Moderate yellowish green | |
| TCS05 | 10 BG 6/4 | Light bluish green | |
| TCS06 | 5 PB 6/8 | Light blue | |
| TCS07 | 2,5 P 6/8 | Light violet | |
| TCS08 | 10 P 6/8 | Light reddish purple | |
| TCS09 | 4,5 R 4/13 | Strong red | |
| TCS10 | 5 Y 8/10 | Strong yellow | |
| TCS11 | 4,5 G 5/8 | Strong green | |
| TCS12 | 3 PB 3/11 | Strong blue | |
| TCS13 | 5 YR 8/4 | Light yellowish pink (skin) | |
| TCS14 | 5 GY 4/4 | Moderate olive green (leaf) | |

Figure 3.14 Approximate Munsell test color samples. (*Source:* http://en.wikipedia .org/wiki/Color_rendering_ index under license http:// creativecommons.org/ licenses/by-sa/3.0/.)

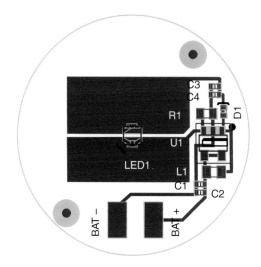

Figure 10.5  Layout of LED flashlight.

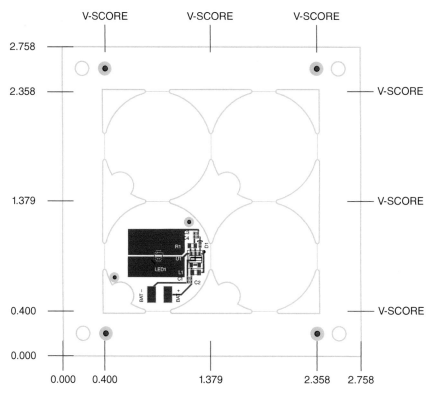

Figure 10.6  Panelization of LED flashlight.

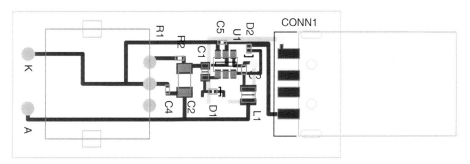

Figure 11.6    USB light PCB, top view and X-ray bottom view.

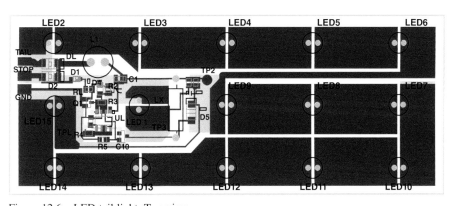

Figure 12.6    LED tail light. Top view.

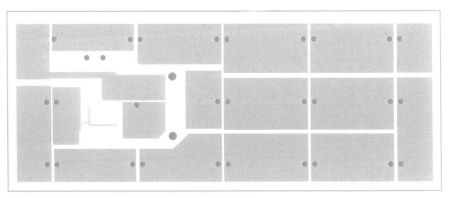

Figure 12.7   LED Tail light. Bottom view.

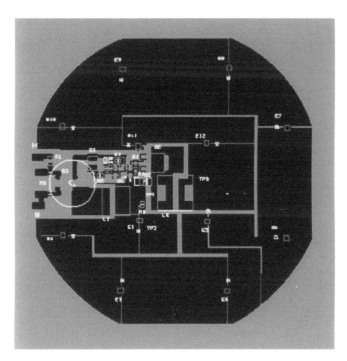

Figure 13.6    BR40 bulb PCB.

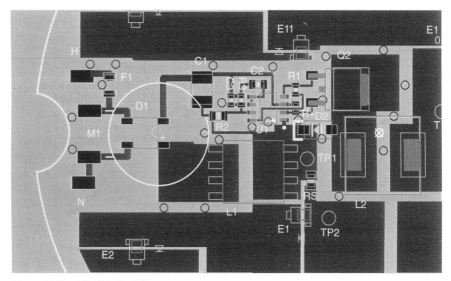

Figure 13.7    BR40 bulb PCB zoom.

# CHAPTER 13

# PRACTICAL DESIGN OF AN LED LIGHT BULB

## INITIAL MARKETING INPUT

The company has decided to move into the field of residential lighting. For their first product, they "want a floodlight light bulb. I (the sales VP is talking) have a 14 ft ceiling in my house, and it's a bear to change out those light bulbs. So I want something that's going to last as long as my house, so I don't ever have to get out the ladder again. I've seen approximately what I want at Home Depot. It's 1260 lumens, 2700 K, and sells for $20. The places we're selling in to expect 40% margin, so we have to sell it to them at $12. And that means I need your cost, including the mechanical, to be $6."

A little discussion about spotlights versus floodlights establishes that he does in fact want a floodlight, so we don't need special optics to control the angle of emission. The discussion then turns to performance trade-offs. The cost is not the only thing limiting the brightness. To make the light bulb brighter, the LEDs are also going to have to run hotter. And hotter LEDs means the MTTF (mean time to failure) is going to be less. Here the answer is that "it has to last longer than a fluorescent tube, say at least 25,000 h. But give me a table of brightness versus life so we can choose."

## INITIAL ANALYSIS

There are going to be two sources of heat in this design, the ballast and the LEDs. Let's take a wild guess that we can put in maybe 10 W into this design. (If we're off 50% either way it won't affect our zero-order conclusions.) If we have 85% efficiency on the ballast, that's about 1.5 W in the ballast, and the rest in the LEDs. Since reflectors are pretty big, we're going to assume that the only real problem is going to be getting the heat out of the LEDs; the ballast isn't going to overheat.

So, how much heat can we really put in? The first thing to note, as usual, is that "floodlight" doesn't really capture all the design information we need. We have to know the size of the bulb to determine the thermals. Looking at the General Electric

*Practical Lighting Design with LEDs*, Second Edition. Ron Lenk and Carol Lenk.
© 2017 by The Institute of Electrical and Electronics Engineers, Inc. Published 2017 by John Wiley & Sons, Inc.

Figure 13.1    BR bulbs.

Lighting web page[1] on indoor floodlights shows that the word "floodlight" encompasses bulb sizes of BR30, BR40, MR16, PAR16, PAR20, PAR30, PAR30L, PAR38, R20, R25, R30, and R40! We narrow this down by looking just at medium screw bases, which somewhat shortens the list of sizes to BR30, BR40, PAR30L, PAR38, R20, R25, R30, and R40 (Fig. 13.1).

Further examination shows that a lot of these sizes are available only for compact fluorescents. We're tempted to say that marketing wants only the incandescent replacement sizes (BR30, BR40, and R20) because the CRI of the fluorescent bulbs is only 82, and he no doubt wants his house to look beautiful. That would mean using high CRI LEDs. Of course, using high CRI LEDs will reduce the achievable brightness. Maybe energy-conscious consumers, the type who will spend extra money for an LED bulb rather than a CFL, already are using CFLs and don't mind the low CRI. Or on the other hand, maybe they're dissatisfied with the CRI of fluorescents, and that's why they want to change to LEDs. We'll stick to high CRI LEDs and see what brightness we end up with.

Marketing states authoritatively that they had the BR style in mind ("they don't have that odd shape the CFLs come in [R]"). We tell them that the brightest will be the largest, and so we agree on the BR40 size bulb. This is apparently what he saw in Home Depot anyway. This still begs the question of CCT. Marketing of course wants the bulb to look "as close to incandescent as you possibly can." After pointing out that we can't get a perfect CRI, we further mention that what they're asking for means a CCT of 2700 K, which will have significant impact on the brightness. But marketing is quite insistent: CCT of 2700 K, CRI of 90, 1260 Lm, and $6 BOM cost. Make it happen. So we'll go ahead with the design as best possible, and then rely on mechanical engineering to ensure the LEDs don't overheat.

First thing then is to locate suitable LEDs. As usual, we go to Digikey. Typing in "LED," we select "LED Lighting – White." Checking "In Stock" gives 3397 choices. Next we select everything in CCT between 2700 and 2750 K, as 3000 K seems too high. This reduces choices by a factor of 10, down to 388. And for the decisive step, we select only parts whose CRI is 90 or more (there's one with CRI 95). This leaves 73 choices.

---

[1] http://genet.gelighting.com/LightProducts/Dispatcher?REQUEST=RESULTPAGE&CHANNEL= Consumer&FILTER=FT0010:Track+%26+Recessed_Indoor+Floodlight&CATEGORY=Lamps& BREADCRUMP=Track+%26+Recessed_Indoor+Floodlight%230

Since we're tightly constrained on the cost, we need to construct a table of $/Lm. This isn't available directly on Digikey, but we don't have to type in price and flux for each part in Excel, we have a shortcut. At the bottom of the web page is a button "Download Table," which produces a CSV (comma separated value) table. Opening this in Excel, we have the data on all the 73 choices. After a little manipulation and adding a formula to divide flux by cost, we get a column showing Lm/$, and we sort from maximum to minimum. Quite astoundingly, we find the highest value to be a Luminus 3020 part at 995 Lm/$! To get 1260 Lm (ignoring optical losses for the moment) would be only $1.27!

Is there a downside? Digikey reports 77 Lm at 120 mA, so we would need 17 devices, not accounting for temperature roll-off of the flux. Can they be run at higher current? A glance at the datasheet shows that absolute maximum current is 150 mA, at which flux is 20% higher. That isn't a huge difference, so for the time being we'll assume 17 LEDs. But that doesn't account for board space, because each LED needs a bunch of vias to keep it cool. And it doesn't account for stuffing costs, the money the manufacturer charges for each component to be put on the board and soldered down. And with 10% optical loss and 10% roll-off with temperature, we're up to 21 LEDs.

Are there other choices? Well, just going down one line in the Lm/$ table, we find the Luminus 3030 part (the previous one was 3020) at virtually the same performance, but generating 104 Lm at 150 mA. And its absolute maximum current is 240 mA, at which it has 42% more flux! Even running at nominal, that would take only 1260 Lm/104 Lm = 12 devices instead of 17. So that's the LED we choose, the MP3030-2200-27-90.

Since we're meeting marketing's flux specification (to first order) at the right CCT and the right CRI, we don't need to construct a table of achievable brightness. Further, lifetime is just set by LM80. As long as we don't exceed that current, the lifetime is (sort of) guaranteed. The one thing left is to make sure the 3030 LED isn't going to overheat. It has an absolute maximum junction temperature of 125 °C. To find what the junction temperature is going to be, what is needed is a thermal model. We'll first model the whole PCB's power dissipation and multiply by the thermal resistance to ambient. That will give us the board temperature. Then we'll look at a single LED's power dissipation and multiply by the thermal resistance to the board, and then add the two pieces together.

Since a BR40 goes into a can, we'd better assume that the light bulb ambient goes up to 50 °C. We're mounting the LEDs and the driver onto a MCPCB, since they are now the same price as FR4 boards. However, we're going to use the lower cost 1 W/(m K) type metal core board. The 3 W/(m K) material has problems with low glass-transition temperature, so we won't be using it. Let's take a quick estimate of the thermal resistance. If the board fits in the top of the bulb, it will have a diameter of about 5 in., which is a board area of $\pi$ (5 in./2)$^2$ = 19.6 in.$^2$ = 0.0127 m$^2$; and assume the board thickness is 0.064 in. = 0.0016 m. Then the thermal resistance through the MCPCB is only 1 W/(m K) × 0.0127 m$^2$/0.0016 m = 7.79 W/K, which is 0.128 K/W.

The PCB will in turn be mounted to a big piece of aluminum, which for our purposes we'll assume to have zero thermal resistance. Finally, there is thermal resistance to the ambient. This depends on a number of factors, such as bulb size versus can size, and whether the can is insulated or not. We can't use the estimates of

Chapter 6 because the can doesn't allow free access for the board to the air for either convection or radiation. We'll take an initial, hopefully conservative, guess of 5 K/W, and measure it later. We thus estimate the total thermal resistance from LED case to ambient to be 0.13 K/W + 5 K/W = 5.13 K/W. This is going to be multiplied by the power the total unit is dissipating.

The second part of the calculation is the thermal resistance of the 3030 package, which the datasheet shows is typically 11 K/W. The LED has nominally 150 mA going through it, and a typical forward voltage of 6.1 V, so it's dissipating 915 mW. This means the junction temperature will be 915 mW × 11 K/W = 10 °C higher than the board.

Now we have all the pieces, see Figure 13.2. The junction can't be more than 125 °C. That means the board can't be more than 115 °C. The difference between that and the ambient is 115 °C − 50 °C = 65 °C. And that goes through a thermal resistance of 5 K/W, so the maximum power is 65 °C/5 K/W = 13 W.

Is that adequate power? We're putting 13 W of power into the unit. Our table shows that at nominal current the 3030's efficacy is 112 Lm/W (at 25 °C!). So that's 1456 Lm, which sounds promising. Let's try doing the calculation the other way around. We need 1260 Lm + 10% temperature roll-off + 10% optical loss + 15% driver loss, that's 1753 Lm. We're getting 112 Lm/W, so we need 1753 Lm/112 Lm/W = 15.65 W. For a 65 °C drop from the board to ambient, that's a thermal resistance of 4.3 K/W. The PCB to metal interface is negligible, so we need 4.3 K/W for the thermal resistance of the outside of the bulb to the ambient. That seems fine, it's still quite close to our initial guess.

Since the power has changed, we'd better go back and check on the number of LEDs. At nominal drive, each LED can produce 104 Lm. If we have 12 of them, that's only 1248 Lm. We said we could overdrive them up to 240 mA, at which current it has 42% more flux. In fact, we only need 1525 Lm/1248 Lm = 22% more flux. Looking at the LED datasheet, we can get that at a drive current of about 190 mA, so we're good to go.

Of course, we could get higher efficacy by dropping the CRI down to 80, and/or increasing the CCT to 3000 K. But marketing was quite specific about not doing that. Can mechanical design ensure that the thermal resistance of the bulb to ambient isn't more than 4.3 K/W? After some discussion, they decide that that can be accomplished by adding fins to the bulb. It doesn't affect the unit cost that much, and marketing expresses the opinion that there are lots of high power bulbs with fins already, so the appearance won't matter that much. Besides, the unit at Home Depot uses 18 W, so we should be able to do it too. Marketing then wants to know whether we are spending extra money to get the efficacy higher. But of course, we're limited by thermal design, so no, we can't make it cheaper by increasing the input power.

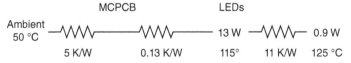

Figure 13.2   Initial thermal model for LEDs in a BR40.

# SPECIFICATION

We thus draw up a specification (Table 13.1), for marketing sign-off before the design begins. There are a number of issues that are indirectly addressed in this specification. In production, we need to hold the tolerance of the light output. Incandescent bulbs used to be almost exactly the right brightness, but LED bulbs of course can't be, with all the tolerances in flux binning, drive current, and so on. On the other hand, the government says that lamps have to be 30% different in brightness for people to perceive a difference. That would be ±15%, so ±10% seems reasonable.

Just as with the tail light in the previous chapter, it's going to have to be verified on the production line by measuring the current through the LEDs. We'll add a 1 $\Omega$ resistor and two test points, the same as for the tail light (power dissipation will only be $[190 \text{ mA}]^2 \times 1 \,\Omega = 36 \text{ mW}$). Of course, the light gets dimmer as it warms, and the production line can't wait for that to happen. What we'll really do is to measure the current at turn on at room temperature, and then apply a correction factor for the efficacy of the LEDs when they're hot. In practice, there will simply be a pass/fail pair of numbers, based on validating the flux with an integrating sphere. If the current is between those two numbers, the light will be the right brightness after it heats up.

The line voltage over which the unit is supposed to operate properly is the standard 85–132 VAC. However, the input voltage is allowed to go as high as 135 VAC and as low as zero. Producing light going down to zero volts is of course not possible. What is meant is that the unit won't be damaged by any line voltage less than nominal. The potential problem is that supplying the LEDs with constant power means the input current has to go up as the input voltage goes down. Many components will dissipate more power as current increases, potentially damaging them. This could in theory be prevented by turning off the ballast when the line voltage goes below some value. But the other way to accomplish this is to have the ballast be power factor corrected. That way, the power goes down as the voltage decreases, so the current doesn't increase.

**TABLE 13.1    Specifications for LED BR40 Light Bulb**

| Parameter | Min. | Typ. | Max. | Units |
|---|---|---|---|---|
| Brightness (by design) | 1134 | 1260 | 1386 | Lumens |
| CCT | | 2700 | | K |
| Operating input voltage | 85 | 120 | 132 | VAC |
| Safe input operating voltage | 0 | | 135 | VAC |
| PF | 0.9 | | | |
| Input power | | 15.7 | | W |
| Efficacy | | 83 | | Lm/W |
| Lightning protection | | Yes | | |
| AC conversion efficiency target | | 90 | | % |
| Triac dimmer compatibility | | Yes | | |
| Ambient temperature (by design) | 10 | | 50 | °C |
| MTTF | | 25,000 | | h |
| Dimensions | | BR40 | | |
| Production cost target | | $6.00 | | US$ |

Since we know how to do this inexpensively (Chapter 8), we decide to include PFC. Marketing likes this anyway, since that means they can get Energy Star rating on the bulb. And decreasing brightness is what incandescent light bulbs do anyway, so they don't expect pushback from customers.

Input power is based on our calculations for the moment. When we get to mass production, we'll add in a minimum and a maximum value based on production line test results. Efficacy is simply calculated from flux and power. It's for marketing purposes and doesn't add anything to the actual specification. Of course the bulb needs to be lightning protected.

We target an AC conversion efficiency of 90%, since we used that for our power calculations. If we can't achieve that for some reason, it may add to the cost to get better components. Hopefully, such a decision won't have to be made. Dimmer compatibility is fairly straightforward, since the power supply works when voltage is present, and doesn't work when it isn't. We do need to specify that dimming is to be on a triac, not 0–10 V. Something to watch out for is that the IC will need to start right away when the voltage appears. Of course, we also need to test to see if the circuit flickers at some point on the dimming curve, but we don't specify how low the dim level has to go. So this will probably be OK.

No typical ambient temperature is specified. There are lots of different environments for the bulb, and it needs to have at least its rated MTTF for all of them. The bulb shape is specified to be a BR40. We discuss some of the options here with marketing before deciding this. There are of course the usual trade-off arguments. On one hand, marketing thinks the bulb should be exactly the same as the incandescent shape, so that consumers don't have to get used to something new. On the other hand, that shape was certainly not designed to optimize the thermal performance of LEDs. Remember that mechanical engineering said they needed fins for heat dissipation. After all, once it's in the ceiling in a can, no one pays any attention to what shape it is anyway. Marketing already agreed to this.

There's an additional consideration here. It might be supposed that the BR40 is so large (5 in. at its widest) that it will block up the possibility of airflow in a 6 in. can. Maybe air flow can be used to increase the power dissipation capability? Unfortunately, this idea doesn't work. Since the can and its lamp are pointing downward, the hot air goes up *into* the can. Air circulation is probably minimal inside the can. Further, the LEDs are presumably at the top, flat end of the bulb (closest to the floor). Thus there really isn't much reason to deviate from the standard shape, and a really good reason not to make it smaller: That would reduce the top surface area, which is the primary site for heat dissipation. One further choice would be a PAR38, which is almost the same size as the BR40, but with a flatter top. If we can somehow shorten the path from the top to the ambient, that would help. We hold this idea in our back pocket.

## POWER CONVERSION

Let's start with our HV9910 design for off-line applications, Figure 13.3. It already includes power factor correction, according to the design in Chapter 8, so we only need a single stage. If we had to have two separate converter stages, one for power

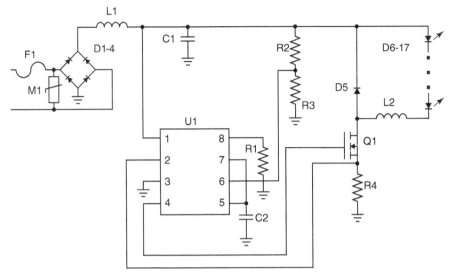

Figure 13.3    Initial design for a BR bulb.

factor correction followed by one to run the LEDs at constant current, we probably couldn't meet either the cost or the efficiency goals.

Let's start right at the beginning of the circuit with the fuse. We're only pulling 15 W from the AC line, so the current at 120 VAC is only 120 mA. Remember that since this circuit is PFC, the current goes down as line voltage goes down. So at maximum voltage of 10% high line, the power is up only 10% voltage × 10% current = 20%. So, for the fuse we can select something with about double that current rating, a 1/4 A fuse. We need inexpensive, which suggests a through-hole. But this is a metal-core PCB, so everything needs to be surface mount. We start looking through Digikey. Selecting "Fuses," "In Stock," "250 mA," and "Surface Mount" leaves 60 choices. We want only those that are rated at least 125 VAC, leaving 36 choices. Sorting by price, the cheapest is from Bel Fuse, a C1F 250. It's in a 1206 package, which is quite small for this board.

We'll look at the MOV and the input filter last. The input filter design is going to depend on our switching frequency, while the MOV is going to be the biggest device that fits. Looking further on, the next components are the diodes D1-4. They are nothing special in terms of speed. As for current, we can choose 1 A devices; there isn't much difference in size or cost to go lower in current. Rather than deal with the stuffing costs of four components, we will instead use an integrated diode bridge. Looking to Digikey again, "Bridge Rectifiers," "In Stock," "Surface Mount," 400–600 V and 1 A leaves 55 choices. The cheapest is the MDB6S, so we'll use that.

Now we can turn to the main part of the circuit. The most critical decision for the rest of the components is the switching frequency of U1, the HV9910. Knowing this allows us to design everything else. We follow the guide shown in Chapter 8 and settle for a switching frequency of just less than 150 kHz. Remember that this is so because the fundamental of the power is below the point where EMI starts to be measured. That way, the first harmonic we will need to filter will be the third, close to 450 kHz.

The tolerance on the switching frequency of the IC is ±20% at 25 °C. Let's add in another 5% for temperature variation, and 1% for the resistor tolerance. To be sure we don't hit 150 kHz, we need to choose the switching frequency to be 150 kHz × (0.8) × (0.95) × (0.99) = 113 kHz. And since the datasheet shows that the IC is apparently tested at 100 kHz, we just round down to this frequency, and use R1 as 226 kΩ. That puts our third harmonic at 300 kHz. While we're at it, let's make the converter constant-off time, which can be done simply by attaching R1 to the gate of Q1 rather than to ground. That will also help the noise, without significantly affecting operating performance.

With the switching frequency in place, we can select the inductor. Following the information in the HV9910 datasheet, we have the duty cycle as a function of input voltage $D = V_{LEDs}/V_{in} = (12 \times 6.3 \text{ V})/V_{in}$. The on-time of the switch is $t_{on} = D/F_S = 75.6 \text{ V}/(V_{in} \times 100 \text{ kHz})$. The recommended inductance is then $L = (V_{in} - V_{LEDs}) \times t_{on}/ (0.3 \times I_{LED}) = (V_{in} - 75.6 \text{ V}) \times 75.6 \text{ V}/(V_{in} \times 100 \text{ kHz})/(0.3 \times 0.19 \text{ A})$. Clearly, the largest inductance would be needed at the highest line voltage. Since this design is PFC, the peak voltage is $120\sqrt{2} = 168 \text{ V}$. Substituting, we find $L = 7.3 \text{ mH}$. Its value is large primarily because the current is so small.

Actually, this calculation has an error in it. Did you spot it? We want 190 mA *average* through the LEDs. But the ballast is PFC, and so the current is a function of the line voltage. In particular, to get 190 mA at the average voltage (120 V) we need 190 mA × $\sqrt{2} = 270$ mA current at line peak. Substituting this corrected current in to the inductance calculation, we find that the actual inductance required is 5.1 mH.

Now, the datasheet and inductance formula assumed that the ripple current was 30% peak-to-peak of the current. At line peak we have 270 mA, and so the peak current is $I + (30\%/2) \times I = 310$ mA. We need to find an L2 that is 5.1 mH with 310 mA saturation current. Selecting "Inductor," "Fixed Inductor," and "In stock" at Digikey as usual gives a huge number of choices. To cover as many devices as possible, we look at inductors between 4.3 and 6.8 mH, yielding 166 choices. And then, selecting saturation currents of at least 300 mA narrows it down to 28 choices. Sorting by price, the least expensive choice is an SRR1208 from Bourns, 4.7 mH with a saturation current of 300 mA. It's surface-mount. Since it's just barely meeting our specs, it must be the best we can do. Power loss will be of $(190 \text{ mA})^2 \times 9.6 \Omega = 347$ mW, which sounds OK. And as it turns out, it's shielded! That will help our radiated EMI.

Knowing the current, we can now select the current sense resistor, R4. At line peak we're going to set the PFC circuit to produce the 250 mV that is full gain for the HV9910. So we want the line peak current in the transistor to be 310 mA. (Remember that the HV9910 trips off at a current level, so we have to include the ripple to make sure we get enough current.) Since the current sense pin trips at 250 mV, we want R4 = 250 mV/310 mA = 806 mΩ. The closest common value is 820 mΩ. To know the power dissipation in R4, we need the duty cycle when it is conducting current. Average duty cycle is $(12 \times 6.3 \text{ V})/120 \text{ V} = 63\%$, and so power is $(190 \text{ mA})^2 \times 0.82 \Omega \times 63\% = 19$ mW, so R4 can be as small as we need.

With knowledge of the current, we can also select Q1 and D5. For the transistor, the square of the average current is only 0.04 A². So we can pick something with several ohms of resistance, since it's only 40 mW for each ohm. On Digikey we select "FETs – Single" and "In Stock." For voltage, we have our normal 400 V of breakdown for semiconductors on 120VAC, but the LEDs are an additional 75 V on top of that, so we will choose voltage of at least 500 V, which still leaves 2855 choices. Selecting a maximum resistance of 4 $\Omega$, the cheapest device is the AOD3N50, a 500 V device with 3 $\Omega$ resistance in a DPAK. There don't seem to be any high-voltage choices that aren't in that package, so we'll have to make it fit. As for gate charge (important because of the limited drive capability of the 9910), it's on the high side at 8 nC. But for 5 ¢ more, we can move to the IPD60R3K3C6, which is 600 V at 3.3 $\Omega$, and only 4.6 nC. We choose this device.

D5 is also carrying the 190 mA average. It's being switched at 100 kHz, so it needs to be ultrafast. Nowadays there are also 600 V schottkies, so we should include those in our search. Going to Digikey and selecting "Diodes, Rectifiers – Single," "In Stock," and at least 400 V gives 2055 choices. We'll select reverse recovery time from 0 to 100 ns, giving 1056 choices. The least expensive in surface mount is an ES1J. Looking at its datasheet, at 200 mA it has a forward voltage of 1.1 V, which would be a power dissipation of 220 mW. For less than a penny more, we can use the RF071, which has a forward voltage at 200 mA of 0.85 V. But it doesn't seem worth it to go to a new device (we've used the ES1J before) just to save 50 mW in a 15 W device, so we stick with the ES1J in an SMA package.

Let's do the PFC circuit next. We want a resistor divider that produces 250 mV when there is 169 V input. At the same time, we don't want to dissipate too much power in the top resistor either. Picking the largest practical value, 1 M$\Omega$, for R2, we need R3 to be 1.48 k$\Omega$. The closest common value is 1.5 k$\Omega$. R2 dissipates only $(120 \text{ V})^2/1 \text{ M}\Omega = 14$ mW. Nonetheless, remember that R2 needs to be in 0805 for voltage breakdown rating.

Some experience suggests that a 1 M$\Omega$ input impedance to a circuit tends to be noisy. So we'll add a capacitor in parallel with R3. With a rectified period of 8.3 ms, we want to track the line voltage with, let's say, not more than 400 µs delay. With the 1 M$\Omega$ resistor, this time constant implies a 400 pF capacitor. We'll round down to the common value 390 pF.

The only components left to design are the input filter components L1 and C1, and the MOV. Turning to the EMI filter first, the input capacitor should not be too large, so that we don't have power factor problems. We limit ourselves to 100 nF, which we find at 630 V, available at reasonable cost with 10% tolerance in an 1812 package.

Given the capacitor, the inductor is determined by how much roll-off we need at the switching frequency. Remember that this is a two-pole filter, since it has both a capacitor and an inductor. That means that the reduction in noise is proportional to the *square* of the frequency, or two times the number of dB. For example, if we place the filter poles at 1/10 the switching frequency, 10 kHz, we get 40 dB of attenuation at the switching frequency. But wait! We can get an additional 12 dB reduction because we went to constant off-time. Doing this, then we get a total of 52 dB reduction at 10 kHz $= 1/(2\pi\sqrt{L} \times 100 \text{ nF})$, so that $L = 2.5$ mH. With input current at 120 mA at 120 VAC, it will be $\sqrt{2}$ higher at line peak, 170 mA. Looking in the range of 2–5 mH and saturation current rating of 180–250 mA, we find 28 choices. Selecting

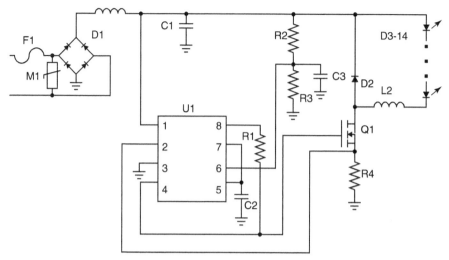

Figure 13.4   Design for a BR bulb.

only shielded ones, since this is after all an EMI filter, we have seven choices, the cheapest being a 3.3 mH from Bourns, an SRR0908. It lists 8 Ω resistance, so power dissipation is only 115 mW.

We've left the MOV for last because it's not determinate. You just pick the biggest one that will fit. But there's a problem here. Surface-mount MOVs really don't work very well, they have very small energy handling capability. Worse, they're something like an order-of-magnitude more cost. But we have a trick that we've used on previous designs. We can have the manufacturer trim the leads of a through-hole component short, and then hand-solder the leads onto the board. This is OK for a single component being produced in China, although you can't do it practically if you have more than one or two components, or for U.S. production. We'll choose a 14 mm disk, and let the layout person complain if it doesn't fit.

The completed schematic is shown in Figure 13.4, with the BOM in Table 13.2. Note that since we are using a dozen LEDs, the cost for 1 K boards uses the pricing for 10,000 LEDs. This reduced the cost per LED even further, to only $0.095. Our BOM cost ended up at $3.64 at 1000 pieces; we can estimate that in volume in China it will be closer to $2. We should be able to make a profit at that even after the mechanicals are added in.

## THD

We've already discussed what power factor is, and how we're going to accomplish power factor correction in this circuit (both in Chapter 8). One thing we haven't talked about yet is total harmonic distortion (THD). While THD has long been familiar to those interested in audio, it's relatively new in the world of power conversion. And while the US government doesn't—yet—mandate a maximum THD on lighting

**TABLE 13.2  BR Bulb BOM**

| Ref. des | Description | Part # | Mfr. | 1 K price ($) |
|---|---|---|---|---|
| C1 | 100 nF, 630 V, 1812 | Any | Any | 0.204 |
| C2 | 100 nF, 10 V, 0805 | Any | Any | 0.017 |
| C3 | 390 pF, 50 V, 0402 | Any | Any | 0.007 |
| D1 | 600 V, 1 A bridge | MDB6S | Fairchild | 0.094 |
| D2 | 600 V, 1 A, ultrafast diode | ES1J | Any | 0.068 |
| F1 | 250 mA, 125 VAC fuse | C1F 250 | Bel fuse | 0.112 |
| L1 | 3.3 mH, 150 mA, 8 Ω | SRR0908-332 | Bourns | 0.476 |
| L2 | 4.7 mH, 280 mA, 9.6 Ω | SRR1208-472 | Bourns | 0.381 |
| M1 | 14 mm, 140 VAC MOV | MOV-14D221 | Bourns | 0.153 |
| Q1 | 600 V, 3.3 W, DPAK | IPD60R3K3C6 | Infineon | 0.269 |
| R1 | 226 kΩ, 1%, 0603 | Any | Any | 0.002 |
| R2 | 1 MΩ, 1%, 0805 | Any | Any | 0.004 |
| R3 | 1.5 kΩ, 1%, 0603 | Any | Any | 0.002 |
| R4 | 820 mΩ, 1%, 0603 | Any | Any | 0.028 |
| R5 | 1 Ω, 1%, 0805 | Any | Any | 0.003 |
| U1 | Controller | HV9910B | Supertex | 0.690 |
| Subtotal | | | | 2.510 |
| LED1-12 | 2700 K, 90 CRI White | MP3030-2200-27-90 | Luminus | 1.145 |
| Total | | | | 3.66 |

products, this is probably a good place to talk about it. So in this section we'll take a look at what THD is. And then we'll do a simplistic analysis of our circuit to get an estimate of the THD we should be expecting.

The basic idea of THD, the same as for PF, is that the current drawn from the AC line should be sinusoidal. And THD is a way of measuring just how nonsinusoidal the current is. Still, it's not interchangeable with PF. Usually, when PF is high (say >0.90, which is good), and THD is low (say <20%, which is also good). But once in a while you come across a circuit that has high PF but also has high THD—so they're not measuring quite the same thing.

Recall, then, that PF is found by taking the integral over a line cycle of the power (that is, the average) divided by the RMS current and the RMS voltage—the RMS also being integrals. So PF is time based, and depends on both the voltage and the current. By contrast, THD is not time based, and depends only on the line current, not the voltage. Specifically, THD finds the Fourier decomposition of the line current waveform, which is to say its harmonic contents. It then computes how much of the current being drawn is in the harmonics of the line frequency (in the U.S. 120, 180, 240 Hz, etc.) versus how much is in the fundamental (60 Hz). As a formula, it's the root of the sum of the squares of the harmonics divided by the fundamental:

$$\text{THD} = \frac{\sqrt{I_{120}^2 + I_{180}^2 + I_{240}^2 + \cdots}}{I_{60}}.$$

Here the sum is over the first 50 harmonics, that is, up to 3 kHz. It's easy to see that when the line current is a pure sine wave at the same frequency as the line voltage,

THD is zero, since there are no harmonics. Usually, good PF and low THD go together. However, you need to notice that you *can* have a good PF while having a bad THD, and vice versa. In particular, PF looks at whether the line current is in phase with the line voltage, while THD only cares about the distortion, not the phase. So, if current is drawn sinusoidally, but out of phase with the line voltage, the THD will be 0%, while the PF may be very low. PF and THD are thus again not measuring quite the same thing.

To get a feel for what kind of THD we may expect from our design, we'll make an approximation. We know that the LEDs won't turn on until the line voltage is high enough to forward bias them. Thus, there will be some time while the line voltage is below threshold that the LEDs are off. After they turn on, the line current will follow the line voltage, and so will be sinusoidal. So let's approximate the line current as a square wave (with duty cycle different than 50%, but that won't matter much). The important part is that the current rises up pretty sharply from zero to the sinusoidal part. It's this edge that gives the dominant contribution to the THD.

In this approximation, then, we can just use a well-known fact about the harmonics of a square wave. Only odd harmonics exist (where $n$ is the number of the harmonic), and each one has an amplitude of $1/n$. So it's easy to figure up the THD:

$$\text{THD} = \frac{\sqrt{\sum_2^{25} \frac{1}{(2n-1)^2}}}{1} = \sqrt{0.2237} = 0.473.$$

Note that the sum goes from 2 to 25 because $(2 \times 2 - 1) = 3$ and $(2 \times 25 - 1) = 49$. We thus estimate that the THD is going to be on the order of 47%—considerably higher than the typical specification of 20%. To bring it down, we would have to add in some form of THD correction. Such a circuit would probably also improve the PF further. But since THD isn't on our specifications, we won't be doing it here.

## FLICKER INDEX AND PERCENT FLICKER

Yet another thing we've touched on in previous chapters is visual flicker of the light. So this would be a good place to discuss the two main current methods of quantifying flicker, Flicker Index and Percent Flicker (which we will abbreviate as FI and %F, respectively). Let's start with a summary in words of how these two measures are calculated. Suppose (as is almost always the case) that the light output varies with time. FI first finds the average light output. It then finds how much of the total light is emitted above that average. FI is then the ratio of the average light to the total light. %F measures the difference between the brightest light emitted and the dimmest light. It divides this difference by the sum of the values of the brightest and dimmest light. In formulaic form,

$$\text{FI} = \frac{\text{Light} > \text{average}}{\text{Total light}},$$

$$\%F = \frac{\text{Max} - \text{Min}}{\text{Max} + \text{Min}} \times 100\%.$$

Analogous to the difference between THD and PF, the difference between FI and %F is that the former is time based—it takes an integral to find the average and the total light—while the latter is just finding two special points in time.

In practice, FI requires some programming to calculate, either on an oscilloscope or in a spreadsheet. %F is very easy to do by hand by making measurements on an oscilloscope. Fortunately, %F is becoming more common. It also turns out that for many designs, typical %F specifications are harder to meet than typical FI specs. In practical design work, we usually just look at %F.

For this particular design, it is easy to estimate both %F and FI. We have already noted that the LEDs are off during the time when the line voltage is below the forward voltage threshold of the LEDs, so the minimum light output during a line cycle is zero. And thus %F = 100%. As for FI, again the light is either on or off (in our approximation) and so the average is just the product of the light when it's on times the duty cycle. We've already estimated the average DC as 63%, so the average light is 63% of the value it has when it's on. Thus the amount of light greater than average is 37%, and thus FI = 0.37. Notice that as expected, the %F is harder to keep low than the FI.

Neither of these flicker measures for this circuit is very good. To fix them would require the addition of some energy storage during the time when the line voltage is low, that is to say, a capacitor. Since the flicker isn't specified in our design, we're not going to add in this complication.

## UTILIZATION

As long as we're looking at all the parameters of the light output from the design, one more measure to consider is utilization. Utilization is how close the LEDs are to producing their maximum rated light output. Utilization is never specified by a customer, it's purely an internal matter. If utilization were 100%, then we would know that we are not spending any extra money on LEDs. This is important because the LEDs are such a significant portion of the total cost of the design. We want to use as few as possible.

In our current design, the LEDs are on at a duty cycle of 63%, and thus this is also our utilization. This means that if we could change the design to one where the LEDs were 100% utilized, we would need 37% fewer LEDs. As we indicated earlier, though, this probably would take a capacitor to cover the time when the line voltage is below the LEDs' forward voltage threshold. In this design, the LEDs cost $1.15, only about 1/3 of the total cost. Reducing the number of LEDs 37% would save $0.43. Since the capacitors are likely to be comparable in cost to this number—and in order to avoid a redesign—we will simply leave the design as is.

## PCB

We're almost ready to turn over the design to layout. But we realize that the one thing we haven't specified yet is the size of the board! A glance at the drawing in

Figure 13.1 shows that there is a fairly large space available for the board. Indeed, if we were to put the board very close to the glass cover, we could have close to 5 in. diameter for the PCB. That would be desirable to get the light-emitting surface to be as large as possible. But on the other hand maybe that's going to cost too much money to make the PC board? We would have board area of $\pi \, (5 \, \text{in.}/2)^2 = 19.6 \, \text{in.}^2$, which at the mass pricing we saw in the last chapter of $0.021/\text{in.}^2$ gives $0.41. That's 11% of our BOM cost—not completely out of line, but it would take our cost up over $4, leaving less than $2 for all the mechanical pieces. Let's instead fix our cost at $4. That means $0.34 for the PCB, which is an area of $16.2 \, \text{in.}^2$, which is a diameter of 4.5 in. Since the bulb is long, we don't expect to have any significant height limitation on components.

The layout engineer produces the schematic shown in Figure 13.5. She said that with a board this big, there won't be any issues fitting things on, and she put the 14 mm MOV disk in the schematic without comment. The same as last time, the LED reference designators have been changed from "D." This time they're "E." Apparently, that's easier to fit on the schematic than "LED."

The layout produced is shown in Figure 13.6. Let's take a close look at this layout. The copper, as usual, is in red, but the background has been set to grey so that the reference designators in yellow are readily visible. On the screen, the layout engineer usually has the background black. The board outline, in green, already has the flat edges for panelization. Note also on the far left the semicircular cutout in the board outline. This is for the power wires to come up to the board. It is sized so that two insulated wires can be readily accommodated. The cutout is actually extra big, in case we want to consider twisting the pair of wires to reduce radiative area for EMI control.

The AC line inputs (marked "H" and "N" for "hot" and "neutral") are on the far left. The MOV is by far the largest component on the board. If the light didn't have a huge neck accommodating a large PCB, we would have had to shrink the MOV down in order to make the design fit. As we discussed, since it's a through-hole component

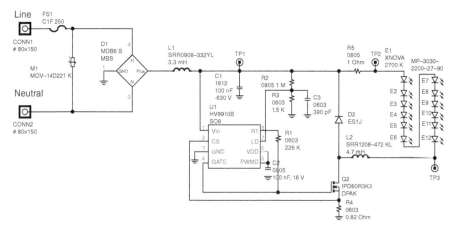

Figure 13.5    Schematic for a BR40 bulb.

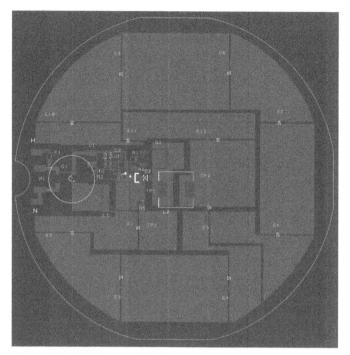

Figure 13.6    BR40 bulb PCB. (See color plate section.)

on a metal-core board, it's hand-soldered down to pads. It's then folded over, in this case on top of the bridge, to avoid any height limitations—or light blockage.

The EMI filter components, L1 and C1, are close to the positive output of the bridge. It would have been nice for L1 to be even closer, but it would bump into the MOV, so it's probably good enough where it ended up. The switching power components, Q2, D2, and L2 are all very closely packed to avoid generating EMI-causing magnetic fields from current loops. And the gate drive, which is often a noise problem in SMPS, has a very short distance to go from pin 4 of U1 to the gate of Q2. It just picks up the timing resistor, R1, on the way.

The current sense resistor, R4, goes exactly across pins 2 and 3 of U1, ensuring a low-noise sense of the current. We've found the noise sensitivity of the IC's current sense to be reduced this way. The only downside is that the timing resistor's connection to the IC goes underneath R4. Since R4 is low impedance, it won't be upset by the small signal going underneath it. Conversely, the timing resistor is *supposed* to pick up the gate drive signal for the spread-spectrum frequency control, so it seems like this will be a happy connection. Taking a look at the ground system, the ground side of R4, which is the power return, goes directly to the ground side of the EMI cap, C1, which is the low-impedance bypass for the power switching. The small-signal grounds, in this design only R2, C2, and C3, connect together with a trace before connecting, through a single trace, to the power ground line. Hooking up the small-signal grounds separately, and with a single connection to the power ground, effectively forms a ground island for signal ground, preventing the IC from getting noise upsets. It's the best we can do on a single-layer board, there's no place for a ground plane. (The metal of the PCB doesn't work as a ground plane because it's electrically isolated from the circuit.)

Finally, the LEDs are approximately evenly distributed around the board. The goal here is to ensure some degree of uniformity of the light output. Of course, it's not perfect. There are 12 LEDs, but the far left wedge of the board is occupied with the SMPS, and so an LED didn't fit there. But E2 and E10 are pretty close, so this will give pretty uniform light, especially with a glass lens on top, and when it's in its intended application up in a ceiling.

## SPACING

An important aspect of layout we haven't discussed yet is spacing. If traces of sufficiently high voltage difference are too close together, current can flow across the gap between the traces, even though they're not touching each other. In blatant cases, components on the board just blow up, and there's some black, charred remains on the board. In cases where there's almost enough spacing, but not quite, you can have corona, which is a low-level current flow. This is more dangerous in some ways, because you don't necessarily notice that anything bad is happening. But over a period of time—and depending on things like humidity—the circuitry may degrade to the point where it fails. There's actually a way to tell if your board is suffering from corona. If you put the board in a very dark room, corona shows up as a somewhat purplish "electrical"-looking mist, like what you see in movies for a mad scientist's equipment, but of course less dramatic.

However, it's of course better to design the layout so that the question of corona doesn't arise. The authors have used the following design rule for years, without ever encountering voltage breakdown. First, we make a voltage map, showing where high voltages are and where low voltages are. Usually, this is not much more than two colored-pencil circles on the schematic. Then during the layout, anytime a high-voltage trace or component comes close to a low-voltage trace or component, we stick a specific-sized circle on the layout (on the documentation layer, so it doesn't show up in silkscreen), and ensure that the component-to-component spacing (or component-to-trace or trace-to-trace) is no closer than the circle permits. The circles then permit the engineer to quickly review whether anything is too close to anything else, without having to pull up the distance-measuring tool.

And what size are the circles? Our rule is to use 55 mils (0.055 in.) between any two points that have a peak of 168 V between them, that is, 120 VAC. If the voltage is less or more, the spacing is proportionally less or more. We are going to use the same spacing to board edge, because the metal of the MCPCB may not be floating, and so there could be a potential between it and this AC circuit as well.

On this board, we have zoomed in on the SMPS section and have turned on the documentation layer as shown in Figure 13.7. The little red circles are each 55 mil in diameter. You can see that the MOV pads, which are closer to the board edge than the input pads, are spaced appropriately. In fact, everything before the bridge is

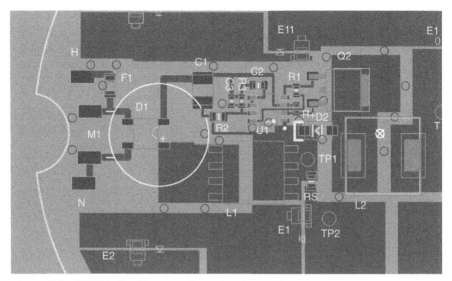

Figure 13.7    BR40 bulb PCB zoom. (See color plate section.)

systematically separated from everything after the bridge by 55 mils, since AC goes both positive and negative.

The plus output of the bridge, as well as the subsequent output of L1, are all shown with separating circles from the low-voltage circuit. Of particular concern for layout is that this high-voltage line also needs to go to pin 1 of U1, to power the IC. So this connection is done with a finger from the copper pour on L1. Thus the spacing circle for R4 (partially obscured by the bar of D2) shows appropriate separation from D2 as well.

Ground at C1, and the grounds hooking up to it, are spaced away from the LEDs. The drain of Q2 and everything attached to it is spaced away from everything else, because the drain switches between zero volts and line peak. Not shown on this zoom, the LEDs also have separation circles between those that are near the top of the string and those near the bottom. Since there is 6 V across each LED, by the time you have 12 of them in series it adds up to 72 V, which requires $(72\,V/168\,V) \times 55\,mils = 24\,mils$ separation. You can see the separations (without the circles) in Figure 13.6, where there are different gaps between different LED copper pours.

## FINAL DESIGN

Our final cost is on the high side, but we can estimate volume pricing by dividing all the costs, except for the LEDs and the IC, by three. This gives $2.44. Added to the $0.34 we estimated for the PCB, we end up with an estimated BOM of $2.78, leaving more than half of the $6 cost target for the mechanical components, assembly, manufacturer profit, and shipping charges. This seems likely to work.

We still need to do several things with this design. Obviously, we need to build a dozen units and verify that they meet the electrical specs. And of course we also need to mount one into the bulb, stick it into a fixture in a thermal chamber, and make sure it doesn't overheat. We need to put one into an integrating sphere to verify light output. We also need to establish appropriate test limits for manufacturing.

But all things considered, this design went pretty well, and given our thorough approach, it will probably work immediately out of the factory, or at least require only minor tweaks. We expect the VP to be pleased, and who knows? Maybe we'll be seeing it in the stores in 6 months.

# CHAPTER 14

# PRACTICAL MEASUREMENT OF LEDs AND LIGHTING

**M**OST OF this book up to this point has been design-oriented: how to calculate how hot the LEDs become, how to design to obtain the desired number of lumens. But by the end of the design cycle, at the latest (but hopefully before), you have to check what is really happening. Maybe the LEDs are running hotter than you expected. Measuring LED temperature is not as simple as just sticking a thermocouple onto one. The light output may be less than it ought to be, or the CRI may be low. You can't just hold a lux meter 3 ft away and tell if it's bright enough, and guessing CRI is unreliable too. This chapter is going to look in detail how to make these sorts of measurements, accurately and in a repeatable fashion.

## MEASURING LIGHT OUTPUT

### Lux Meter

The cheap and easy way to measure the light output of a system is by using a lux meter. A lux meter measures the luminous intensity on a surface. Figure 14.1 shows a typical unit, a Mastech MS6610. It has a detector, the white circular area, which is the sensor part. In a good unit this detector will be "cosine corrected," compensating for the angle that the light is coming in to the sensor. The meter should also be calibrated, so that its measurement takes into account the CIE human eye response curve.

Now, while this device is easy to use, you have to be cautious with this meter. It is very sensitive to the distance from the light source. Remember that lux is proportional to the square of the distance from the light source. Suppose that at 1 m away you measure 3500 lux. If for your next measurement you are 1 cm further away, the measurement will be reduced to $3500 \, \text{lux}/(1.01 \, \text{m}/1 \, \text{m})^2 = 3431$ lux. If these two measurements are taken on two consecutive days, you might conclude that your light source has gotten dimmer. The same problem occurs if you are slightly off-center. The cosine correction fixes the light coming in at an angle to the sensor, but the illuminance actually *is* different when you're off-center. You don't want to come to

*Practical Lighting Design with LEDs*, Second Edition. Ron Lenk and Carol Lenk.
© 2017 by The Institute of Electrical and Electronics Engineers, Inc. Published 2017 by John Wiley & Sons, Inc.

Figure 14.1    Typical lux meter. (*Source*: http://www
.p-mastech.com/products/06_ifep/ms6610.html.)

the conclusion that your light source has dimmed (or gotten brighter) because you varied the measurement setup.

To use this kind of meter, we recommend a setup that fixes the measuring distance. This could be as simple as using a 3 ft cardboard mailing tube: Place the tube over the light source and the lux meter sensor at the other open end of the tube. This eliminates most of the distance variation in measurement. To eliminate angular variation, try making this tube into a semipermanent fixture. Glue it down to a piece of cardboard, and mark the light source position on the cardboard for repeatability. You could also mount the sensor to a second piece of cardboard, with a ring on the cardboard showing where to position it on the tube.

When using a lux meter, one must be careful to have no background light. Also be aware that if a tube or enclosed box is used, the sides of the box or tube will reflect the light and add to the measured numbers. If you're just taking a relative measurement, such as in the case of percentage light drop over a certain time period, then the effects of the sidewalls do not matter. If, however, the actual luminous intensity is to be measured, you should either use a very large box with dark walls or you should line the walls with nonreflective flocking paper. The edges of the box or tube are a source of light leaking in and should have a drape of black material to cover them. A dark measurement, that is, one that is taken with the light source turned off, should be made to determine the background light level. You should do this every time you take a measurement, and subtract it from the number you have measured.

It is tempting to take a lux measurement and multiply by the surface area of an equivalent sphere to calculate the luminous flux. If the light source is uniform, then the equivalent lumens can be calculated this way. However, very few light sources are uniform, including incandescent bulbs. If the light source is uniform but only over a limited angle such as in a reflector lamp, then calculating lumens might give a grossly incorrect answer. In almost all cases, whether the light comes from an LED or from an incandescent light bulb, the light source is not uniform. Therefore, a lux meter cannot

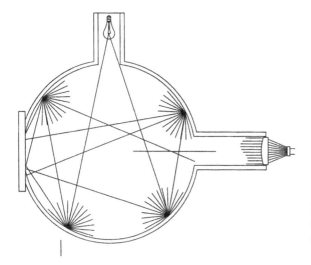

Figure 14.2   Sketch of operation of an integrating sphere. (*Source*: http://www .aaccuracy.com/optical.htm.)

be used to determine lumens. Lux meters are properly used for measuring illuminance, that is, light on an area, and not luminous flux.

## Integrating Sphere

To properly measure luminous flux, an integrating sphere must be used. This is a sphere, preferably four to five times larger in diameter than the light source, into which the light source is placed. They are available in sizes ranging from a few centimeters to those that are big enough for a person to walk into. The inside of the sphere is coated with a highly (>98%) reflective material. Light from the source emitting in any direction is reflected inside the sphere and is measured at a collector (see Figs. 14.2 and 14.3). Thus, the luminous flux is physically integrated across all directions, hence

Figure 14.3   Walk-in integrating sphere.

the name. One can see this effect by peering in a thin crack after the sphere is closed. The inside of the sphere seems to emanate a uniform glow.

It's important that light from your source doesn't shine directly into the integrating sphere's collector. The collector is behind a baffle, a light-blocking element that keeps light from directly reaching the collector. Baffles are designed to be of a minimum size. Larger baffles have a negative effect on sphere accuracy. Care should be taken to ensure that the light source does not leak around the baffle. It is difficult to see leakage by looking for shadows of the baffle created by the light source, since the sphere is so highly reflective. A piece of paper held behind the baffle shows the shadows more clearly. If the shadow does not completely cover the collector, then the light source should be moved until the baffle completely covers the collector. In the case of a directional light, the light source may be pointed away from the collector. When all else fails, a quick fix is to tape a piece of white paper, as small as possible, to the stage (mount) of the light source to aid the baffle.

The integrating sphere is calibrated to a standard light source, a tungsten light bulb. The standard should be an NIST traceable standard, which means that it is calibrated against another standard that is calibrated to a standard at NIST with a known lumen output. The calibration is said to be transferred from the NIST standard to the standard at a lab, and then transferred to the standard that you buy. Calibration of the sphere will last for a year or for 50 h of actual on-time, whichever is reached first.

A calibration bulb comes with a calibration file that relates how that standard compares with the vendor's standard. At your lab, you will transfer the calibration from your standard to an auxiliary bulb that stays permanently inside the sphere. The calibration bulb with a known luminous flux is placed on the stage inside the sphere. A high precision current source is used to drive both the calibration bulb and the auxiliary. The calibration bulb is measured first. The amount of light measured at the collector reflects the sensitivity response of the integrating sphere. Then the auxiliary bulb is measured and a luminous flux value can be assigned to it, based on its output compared with the output of the calibration bulb. This "calibration transfer" should be done once a month to maintain the auxiliary bulb calibration. The auxiliary bulb is the daily workhorse and performs the absorption correction for the sphere. That way your expensive calibration bulb doesn't need frequent replacement.

When a light source is placed inside the integrating sphere, it changes the system's response. In other words, the surface of the light source is not 100% reflective and it will absorb some light and affect the calibration, and in turn the measurement. This self-absorption is corrected by turning off the light source under test and turning on the auxiliary light. A measurement now will show the effect of the absorption and can be added to the calibration. After all this setup, the unit under test can finally be measured. It's best to follow this proper procedure every time, to eliminate questions about measurement accuracy or drift. It seems like a lot of work, but it becomes routine with time.

The collector of the integrating sphere can be connected to any light collecting meter. The preferred kind is the spectroradiometer (see Fig. 14.4). This meter separates the light by wavelength bins and measures the amount of light at each wavelength. The resulting spectral distribution is necessary for calculations of CCT,

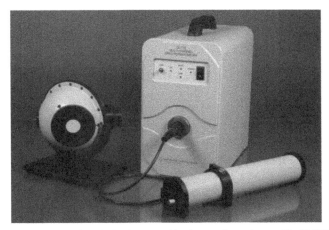

Figure 14.4   Gooch & Housego 6 in. integrating sphere with OL770-LED spectroradiometer. (*Source*: http://www.ghinstruments.com/products/spectroradiometers/ol-770-led-test-and-measurement-system/.)

chromaticity coordinates, and CRI. Alternatively, a photometer can be connected to the integrating sphere. This would be similar to the lux meter, able to measure only lumens and not the spectral information, CCT, chromaticity coordinates, or CRI. In fact, a photometer is basically a lux meter measuring the irradiance at the collector port. The difference is that the integrating sphere reflects the light evenly inside and allows the irradiance to be converted to flux.

Standard sphere size for most labs is 20 in. to measure light bulb size sources. Spheres measuring 6 in. are fine for measuring LED-sized sources. The surface area of the light source should be less than 2% of the surface area of the sphere. A general rule is that the diameter of the sphere should be at least four times larger than the diameter of the light source. The 6 in. spheres have a port against which to place the light source. These are referred to as $2\pi$ measurements since light is being emitted only from the front side of the source and only half of the solid angle needs to be measured. To measure light bulbs inside recessed can lights, a 6-in. port may be added to a 20-in. sphere to perform the same $2\pi$ measurement. If the recessed light is only 4 in. in diameter, a white cover plate should be used to cover up the gap.

## Goniophotometer

Measuring angular light distribution is important for bulbs such as floodlights and spotlights that have prescribed beam angles. To measure angular light distribution, a goniometer is used. When combined with the light measurement device, the system is called a goniophotometer. These are room-sized contraptions with a sensor mounted on a swing arm. Because of the size and cost of this process, these measurements are usually performed in testing labs.

One can obtain an approximate measurement by using a 6 in. integrating sphere to measure irradiance. The light source is mounted on a rotation stage at a distance at least five times that of the largest dimension of the light source. Care is taken to ensure that no other light sources are interfering. Then, relative light distribution can be measured by turning the light source.

## SPECIAL CONSIDERATIONS IN MEASURING LIGHT OUTPUT OF LEDs

Measuring light output of LEDs requires special considerations beyond those for measuring end-product light sources. The light-capturing aspect is straightforward; a 6 in. integrating sphere will suffice. A larger sphere also may be used as long as it is sensitive enough to accurately measure the amount of light put out by a single LED.

The tricky part is due to the self-heating of the LEDs. This leads to continually dropping lumens until thermal equilibrium is reached. Even if the LED is mounted on a large heat sink, the junction will likely rise 10–20 °C in the second that it takes to make the measurement. This can result in a difference of several percentage points in luminous flux from a measurement taken at room temperature. LED manufacturers use a 25 ms pulse of current to the LED to ensure that the junction temperature remains at room temperature. Short of using special equipment, one is unlikely to obtain the same luminous flux measurement as that shown on data sheets. An alternative method has been proposed by NIST (Zhong and Ohno, 2008) wherein the LED is mounted on an electric cooler and allowed to settle to steady state before a measurement is taken.

## LED MEASUREMENT STANDARDS

### Luminaire Light Output (LM-79)

The Illuminating Engineering Society of North America (IESNA) has a standard, IES LM-79-2008, specifying that the light output of LED-based luminaires must be measured inside an integrating sphere at thermal steady state. It reflects the best practices in light measurement and prevents ill-founded or unscrupulous claims of light output. The old practice with incandescent bulbs was to use a relative method of taking the bulb light output and multiplying it by the optical efficiency of the rest of the fixture. With LED-based luminaires, the fixture design is integral to the thermal performance and would affect the light output in terms of both optical loss and thermal roll-off. The net light output must be measured in an integrating sphere, the absolute method. LM-79 also specifies that the power into the driver for the LEDs is the net input power. Net light output divided by the net input power gives the luminaire efficacy.

### LED Lifetime (LM-80)

The other IESNA standard that concerns us is LM-80-2008. The intent of this is to standardize LED manufacturer lifetime testing procedures. As was mentioned before, LEDs don't burn out, they just get dimmer over time. LM-80 defines LED end of life as the time when the LED reaches 70% of its initial lumens ($L_{70}$). It also specifies that the LEDs should be tested for 6000 h (8.3 months) at three different junction temperatures.

However, LM-80 stops short of specifying how to extrapolate this information to a useful lifetime, because there isn't a definitive model yet. So besides verifying that the LED manufacturer follows LM-80 lifetime testing guidelines, one should also verify that the manufacturer has measured dimming for a reasonable amount of time to allow reasonable extrapolation to lifetime. For example, 1000 h of testing should not lead to claims of a 50,000 h lifetime.

The lifetime issue becomes important when qualifying for Energy Star certification. For that, you must use LEDs from a manufacturer that has LM-80 compliant test data. You must measure the LED temperature following the manufacturer's recommendation and verify that it is within the LM-80 test temperatures. Then the LM-80 data must be greater than 91.8% of initial lumens after 6000 h of operation to qualify for a 25,000 h life rating. It must be greater than 94.1% for a 35,000 h life rating.

## ASSIST

The Alliance for Solid-State Illumination Systems and Technologies (ASSIST), at the Lighting Research Center at Rensselaer Polytechnic Institute in New York, has put together a number of recommended test procedures for LED lights, focusing on tests that emulate real-life operation. A current list of such recommendations can be found at http://www.lrc.rpi.edu/programs/solidstate/assist/recommends.asp. Particularly important is "LED Life for General Lighting," which recommends 6000 h of testing for LED lights. They recommend then "a functional fit (e.g., exponential decay) to the data" in order to predict the lifetime of the light.

## MEASURING LED TEMPERATURE

Recall that LEDs are current devices. You can't set their optical output with a voltage. The optical output depends on the current. Furthermore, a small change in the voltage will result in a large change in the current. Therefore, as discussed in Chapter 7, control of the LED requires a current. This current is often a constant with time.

Diodes generically change their forward voltage with temperature. This is true of LEDs as well. As the temperature increases, the forward voltage of the LED decreases. However, fully understanding this concept requires some more work.

Think back to the introductory chapter and recall that the LED consists of a semiconductor die, some phosphor, and a package. Of these, the electrical characteristics are determined exclusively by the semiconductor die. When we talk about the temperature affecting the forward voltage, we are necessarily talking about the die temperature.

Unfortunately, the die is inside the package, and its temperature cannot be directly measured. Think of how you might try to measure the die temperature. The

Figure 14.5   Location of thermo-couple to measure LED temperature. (*Source*: Thermal Design Using Luxeon® Power Light Sources, Lumileds AB05, 2006.)

standard method for measuring temperature is to attach a thermocouple and use a thermometer. This won't work in this case for at least two reasons. For one, the die is an electrical circuit with very tiny elements. A thermocouple consists of two pieces of metal, and the metal might short out the circuit elements. In addition, since the die is very small, the added mass of the thermocouple would change the thermal perform-ance of the LED. In any case, attaching a thermocouple requires opening up the package of the LED, which is destructive to any other tests you might be trying to run on the device.

Another method of measuring the die temperature is with an infrared (IR) thermometer. While this is routine with normal semiconductors, it might be a problem with a source emitting intense light. And again, this would involve opening the package of the LED.

There are two standard methods of measuring the die temperature of an LED. Both have some problems. The easier of the two methods is to measure the LED package temperature, and infer the die temperature. In the example shown in Figure 14.5, a thermocouple is attached to the package. Simultaneously, the power into the LED is measured. The die temperature is inferred from the equation

$$T_{\text{DIE}} = T_{\text{PKG}} + P \cdot \Theta_{\text{JC}}.$$

Of course, you have to hope that the manufacturer's number for the thermal resistance from the junction to the case, $\Theta_{\text{JC}}$, is correct.

There are a number of practical considerations to note before making this measurement. To start with, the size of the thermocouple should be as small as possible. This helps to avoid any chance of the measurement changing the results. In practice, 36 gauge wire works reasonably well.

Next, the position of the thermocouple is quite important. The better manufac-turers will specify the point at which the package is to be measured. Failing this, attach the thermocouple right next to the anode or cathode metal of the device. (Caution! Don't let the thermocouple actually contact the metal. If it does, it will have the same voltage as the LED has. This may affect the meter reading, or may actually be hazardous if the LED is connected to the AC line.)

Attaching the thermocouple properly is something of an art. You shouldn't just place it against the device. This results in poor contact and gives temperature measurements that are artificially low. You certainly shouldn't hold it in place with your finger. The added thermal mass of your finger completely changes the results. Taping the thermocouple down is better, but it is annoying when the adhesive on the tape comes loose with temperature (you might try Kapton tape). A drop of glue is best. Use the smallest drop you can manage, to avoid the thermal mass issue. Use fast-setting glue, such as a 5 min epoxy or cyanoacrylate (Superglue). Also ensure that the bead of the thermocouple and the first 3 mm of wire are flush against the surface.

There is a second method for measuring the die temperature. Since forward voltage is temperature-dependent, you could use the forward voltage to measure the die temperature. In fact, this is what is done for microprocessors. The microprocessor has a diode inside it. This diode is run at the same current as another, external diode of the same type. By measuring the forward voltage of both diodes, the difference in temperature can be computed.

In theory, this same method could be used for LEDs. You could measure the forward voltage of the LED when it turns on and again when it is in steady state, and the difference will be the temperature increase. Unfortunately, it is that first step that is hard. Since LEDs are tiny, they have tiny thermal time constants. Unless you have the equipment to measure the forward voltage very fast, say within the first 25 ms of turning it on, you don't know the initial die temperature. In practical applications, that makes this method unusable for absolute temperature measurement, although it works reasonably well for differential temperature measurements.

## MEASURING THERMAL RESISTANCE

Having measured the temperature of the LED, we can measure the temperatures of other parts of the system using similar methods. For example, you will almost certainly want to measure both the temperature of the outer case of your light and the ambient temperature. So now, knowing the difference in temperature between two points, and the power dissipated, it's straightforward to calculate the thermal resistance of that path, because thermal resistance equals the temperature difference divided by the power.

Not so fast! We still don't know the power in the LEDs. We have accurately measured the electrical power going in to them, but only a part of that is then lost as heat power. The rest is emitted as optical power (there's also some $IR$ power, but it is usually negligible for visible light LEDs). So to accurately calculate the thermal resistance, we need to know how much of the electrical input power is being emitted as heat power.

The wrong way to go about this is to try to look up the optical conversion efficiency of the LED. You get bogged down in all sorts of different numbers and differing measurement techniques by different people, and in the end you're confused. But you have actually already measured all the numbers you need. The spectroradiometer measured the optical power in each wavelength bin, and if you look for it you will find the total optical power in the data. The correct power to use for the thermal resistance, then, is the total electrical power minus the optical power (in watts, of course):

$$\Theta = \frac{\Delta T}{P_{in} - P_{optical}}.$$

Our cautions about accuracy and precision apply to this calculation (see below). You are subtracting two almost equal numbers, and then dividing into a third number. Be careful. You should cross-measure the two thermocouples at a single point to verify that they give the same reading. Since for the measurement of thermal resistance they are being subtracted from one another, the absolute accuracy is not so critical. If one thermocouple reads 35.4 °C and the other reads 34.4 °C, the temperature difference is only 1.0 °C, which gives you only two digits of precision for the difference. This limits your thermal resistance precision also to two digits, even though your temperature probe has three digits displayed. If you put 10 times more power into the system, you could get an additional digit, but in practice that probably won't work. Two digits accuracy is good enough anyway.

If you decide to determine the thermal capacitance, the same cautions apply again. You measure time (presumably) in seconds, so if you reach 63% of steady-state in 10 s, you can't know the thermal capacitance to more than two digits. Slower systems are easier, although again you probably don't need a lot of precision for thermal capacitance.

The one difficult aspect of measuring thermal resistance is the thermal resistance to ambient. A glance back at Figure 6.5 shows that the effective thermal resistance, which is the $y$-axis value divided by the $x$-axis value, is fairly nonlinear in power, and nonlinear again in ambient temperature. The practical thing to do here is to measure the temperature of the case of your light and the temperature of the ambient as close as possible to the actual operating environment as you can. For example, measure a floodlight in a can, not in open space. If you now increase or decrease the input power level (say) by a small amount, you will be able to use the effective thermal resistance as an estimate of what the new temperatures are going to be. Ultimately, of course, there's no substitute for real measurements of temperature.

## MEASURING POWER, POWER FACTOR, AND EFFICIENCY

In addition to optical and thermal measurements, electrical measurements of lights also have to be taken. You need to measure the amount of power going into the light to put on the label, and in order to calculate efficacy. You also need to measure power out of the ballast, to measure its efficiency. And you have to measure power factor, again for labeling purposes. None of these measurements are especially difficult once you know how to take them. It's just that there are many wrong ways to do it that you need to avoid.

## Accuracy versus Precision

Before getting started, some important terminology needs to be discussed. In conversation, we often use accuracy and precision as synonyms, but they're not. The accuracy of a meter, for example, is whether the number it is showing you is correct. Suppose that your meter says the voltage is 5.011 V. If you measured that same voltage with a calibrated standard and it also said 5.011 V, then your meter is accurate to four decimal places. But suppose instead that the calibrated standard says the voltage is really 5.013 V. Then your meter measurement is accurate to only three decimal places. The fourth decimal place isn't right. Writing down that fourth decimal place in your notebook gives false data. Using instead a meter with six decimal places probably gives you better accuracy, but you don't know that unless it's been calibrated.

Precision is reflected in the number of decimal places displayed. In the example above, the precision is four decimal places. The six decimal place display meter has greater precision than the four decimal place display meter. But remember, that doesn't necessarily mean that the accuracy is any better. Both accuracy and precision are important, but in different ways. You need both, especially when you measure efficiency, as discussed in the following section.

## Measuring DC Power

In a DC power system, power is "just" current times voltage. As the scare quotes signify, however, this is not a trivial measurement to take. To start with, voltage has to be measured at some particular place. For example, suppose you're building a driver to run an LED from a USB port, as in Chapter 11. As a test, you hook up the driver to your 5 V laboratory supply. If you now read the voltage from the front panel meter of the supply, there are several problems. To start with, how accurate is the front panel meter? Chances are that no one bothered to calibrate a bench meter, because it's intended to be just an indicator. It's not intended for accuracy. It also probably has low precision. So you need to use a dedicated digital voltmeter (DVM). Of course, the DVM ought to be calibrated. If you measure the voltage today as 5.011 V and tomorrow it's 5.013 V, has the power supply changed, or is there a meter problem?

Next you need to consider where you're going to put the DVM leads. If you hook them up to the terminals of the power supply, you're reading the power supply voltage. This may not be the same as the voltage at the input pins of your device. If there is significant resistance in the wires, or significant current to the device, there will be an IR drop along both the positive and the negative wires. This will make the voltage applied to the device lower than that being sourced by the supply. As an example, suppose you have 1-m-long cables for the positive and negative, and the device is drawing 40 mA. Using a standard 24AWG wire for the cabling, we have a resistance of 84 m$\Omega$/m. With one meter for each polarity, we have a total of 168 m$\Omega$. At 400 mA, this corresponds to a drop of 67 mV between the supply and the device. At 5 V, that's an error of more than 1%. The right thing to do is to put the DVM probes as close as possible to the inputs of the device you're measuring, both positive and negative.

Here is another thing to watch out for. Sometimes you only have one meter, so you measure the voltage, and then disconnect the probes so you can use the same meter to measure the current. This has a problem, because the ammeter drops some voltage; that's how it measures current. So you need to leave the voltmeter in place and find a second meter to measure the current. And again for the same reason, the voltmeter's probes should be closer to the device being tested than the ammeter.

Another problem with disconnecting the voltmeter is that the voltage may change with load current. Most power supplies won't change very much, but they all have finite output impedance. Now, if you turn the dial on the power supply, of course the output voltage changes. More subtly, if you are running from a battery the battery voltage changes all the time. If you change the load current coming out of the battery, its output voltage will change. If you run the battery for more than a minute, it discharges enough that its voltage changes. Again, the conclusion is that you have to have a voltmeter monitoring the voltage continuously.

Now you are measuring the voltage and the current into the device. Power is the product of the two, right? Yes and no. As an example, suppose you've measured the voltage to be 5.011 V and the current to be 252 mA. It's easy to whip out your calculator and find that the product of the two is 1.262772 W. But that answer is *not right*. Your measurement of the voltage was accurate to four decimal places, and your measurement of the current was accurate to only three decimal places. Your power number can't be accurate to seven places when your measurement is only accurate to three! The rule to follow here is that the product of the two measurements can be no more accurate than the least accurate of the individual measurements. In this case, the current and therefore the power are only known to three decimals places. The power is 1.26 W. This is the right number to record in your notebook. If this isn't accurate enough (and it may well not be when you're measuring efficiency) then you need to increase the accuracy of the current measurement.

A brief word about rounding and truncation is in place here. If your calculator shows you 1.2685 W, truncating this number means just dropping the numbers after the significant digits. In this case, truncating to three significant figures gives 1.26 W. Rounding, on the other hand, means that you change the last digit to the nearest number, based on the following digits. In this case, the six gets rounded up to seven, because the number 68 is closer to 70 than it is to 60. For all of the measurements you take, you should be rounding instead of truncating. This power should be recorded as 1.27 W.

Let's turn to some other problems commonly encountered in making electrical measurements. A very common one is that the numbers on the DVMs are jittery. While you're watching, the voltage changes from 5.011 to 5.013 V, then to 5.010 V and then to 5.012 V. There's no pattern. You could write down just some random number from this set, but that doesn't give you that fourth decimal place of precision.

While there are several possible causes for this, the most common is that your meter is being confused by noise. Your SMPS is switching on and off at 45 kHz or

1.6 MHz, and this pulls current at the same frequency. It's almost certain that your input filter still leaves some degree of ripple on the input. This ripple voltage can upset your DVM. At frequencies such as 45 kHz, the DVM may alias the ripple into the measurement, resulting in what appears to be random fluctuations. At frequencies such as 1.6 MHz, although the meter has no response, the ripple may be rectified by the meter's input circuitry, resulting in an offset that may fluctuate with time.

The solution to the problem of jittery numbers is to eliminate the ripple. An easy way to do this is to attach a bunch of capacitance at the input. Of course, this is not for production. The capacitance you add is only to enable accurate measurements. For example, try adding 10 times the amount of input capacitance you would normally use. If your board has a 10 µF capacitor, add 100 µF. Then add some high frequency bypass capacitors as well. Add a 1 µF ceramic in parallel together with a 100nF ceramic. This usually cleans up the jitter problem. If not, there may be more subtle issues involved, but it goes beyond the scope of this book to attempt to detail them and their fixes.

A similar but unrelated problem is temperature drift. When you turn the supply on, the current is 145 mA, but then it starts drifting up to 148, 152 mA, and so on. The problem here is that the device being tested is warming up. As SMPS warm up, their reference voltages change, $IR$ drops change, and so on. What you do about this depends on how accurate your reading needs to be. An easy but low accuracy fix is to make the measurement the moment you turn on the power. If the drift happens over a period of minutes, this may be good enough to make repeatable comparisons between versions of the supply. If you need greater accuracy, or if the drift is quick, you have no choice, you need to let the unit thermally stabilize before you take a measurement. This may take quite a bit of time, but has the advantage that the measurement will be taken under real operating conditions.

We should also briefly talk about meter ranges. When you're drawing 190 mA, it's natural to put the meter on the 200 mA scale rather than the 2 A scale, so you get more significant digits. Then, when you increase the load to 210 mA, you switch the scale up, or it switches automatically. But how do you know that you get the same reading on the two different scales? They need not have exactly the same gain. It could be that 190 on the 200 mA scale reads 0.21 A on the 2 A scale. Potentially, this makes a graph of current have a spurious jump in it. In short, if you're going to be making measurements that are close to the range transition of a meter, you should set the meter to stay fixed at the higher range.

## Measuring AC Power

To start off with, we want to remind you that AC power can be very dangerous. Please make sure you've read the section in Chapter 8 on safety before you start taking measurements on the AC line.

Measuring AC power is considerably more difficult than measuring DC power. This is because the AC voltage and current are constantly changing, going positive and negative, and their voltage and current waveforms need not match up with each other. The fact that they do not match up is related to power factor, which we'll discuss in a moment. The alternating and differing waveforms are related to the power.

It's relatively straightforward to measure the root mean square (RMS) AC voltage, or the RMS AC current. This is the proper computation for AC measurements, because the square of the voltage or current is what dissipates power ($V^2/R$ or $I^2R$). You square the voltage or current, take its average over time, and then take the square root of this number to get back to the right units:

$$V_{rms} = \sqrt{\frac{1}{T} \int_0^T V^2(t) dt}.$$

To measure AC voltage or current, you should use a meter that's rated for that purpose. Since AC line is usually 50 or 60 Hz, meters are typically prepared to measure this frequency (however, a typical meter will not measure the AC voltage or current at high frequency, such as the current through the input capacitor of your SMPS). The difficulty with these meters comes about when you have a SMPS. The SMPS may have a very high crest factor, beyond the rating of your meter (see Fig. 14.6). The crest factor is the ratio of the peak value to the RMS value of the waveform. For a sine wave, such as the AC line voltage, the crest factor is $\sqrt{2} = 1.4$. But the current in Figure 14.6 has a much higher crest factor, maybe 2.0. The high peak current is caused by a large capacitor right after the bridge rectifier. The capacitor's voltage drops only a little bit during the 8.3 ms cycle, and so it charges back up only at line peak. All of the energy the SMPS used during the rest of the 8.3 ms has to be put back into the capacitor during this short time.

Many meters cannot accurately measure RMS values when the crest factor exceeds 1.7. So if you have a SMPS with a large input capacitor (i.e., without power factor correction), you need to find a meter that will handle the crest factor in order to accurately measure the RMS current. (RMS voltage is usually, but not always, easier to measure, because the AC line source impedance is usually, but not always, much lower than the load impedance.)

We now can measure RMS current and voltage. You might suppose that RMS power is the product of these two, but it isn't! The trouble is that the integral of the product may not equal the product of the integrals:

$$P_{rms} = \sqrt{\frac{1}{T} \int_0^T [V(t)I(t)]^2 dt} \neq \sqrt{\frac{1}{T} \int_0^T V^2(t) dt} \cdot \sqrt{\frac{1}{T} \int_0^T I^2(t) dt}.$$

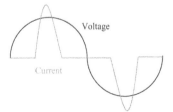
Voltage

Current

Figure 14.6   High current drawn at AC line peak gives a high crest factor.

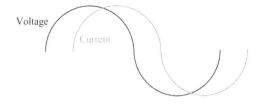

Figure 14.7    Phase shift between current and voltage also gives low power factor.

Without getting into complicated mathematics, let's jump to the answer. Proper measurement of AC power requires a special meter designed for that purpose. The one we have successfully used is a Yokogawa WT210. It simultaneously measures RMS current, voltage, and power. It also reads power factor. It's a little pricey, but we think an instrument similar to this one is the only practical way to take real measurements of AC power.

## Measuring Power Factor

We mentioned that power factor relates to way that the voltage and current waveforms do not match up. The spike in the current waveform of Figure 14.6 certainly gives rise to a poor power factor. But any lag or lead in the current may also result in poor PF (see Fig. 14.7). Again, a suitable meter is requisite for obtaining the right measurement.

## Measuring Ballast Efficiency

Efficiency is the ratio of output power to input power, and, as per by the Second Law of Thermodynamics, is always less than 1. You've seen in Chapters 10–13 that we always have to make assumptions about the efficiency of a ballast during the initial design of an LED light. So once the ballast has been built, we have to verify its efficiency with a measurement.

Measuring the efficiency is as easy as just measuring the input power and the output power—if the efficiency is low, say less than 85%. Of course you need several meters, but that's to be expected. The tricky thing comes about if your ballast efficiency is high. If you measure 10.0 W in and 9.0 W out, the efficiency is 90%. But if the input meter is off by 1%, you get 9.0 W/9.9 W = 91%. Is that 1% due to the new transistor you put in or a meter difference? You can't tell.

The way to avoid questions like this is to take what we call "cross-measurements." Before you take efficiency measurements, put both voltmeters on the same exact node, and verify that they read the same number. It doesn't have to be the *right* number in the sense of absolute accuracy; it just has to be the *same* number. If they both have the same error, it will cancel out when you do the division to determine the efficiency. Do the same thing for the current. Take both ammeters and place them in series with the same exact line. If they both have the same reading, then you will be able to rely on your power measurement. You can do the same if you're using integrated power meters, such as the WT210 mentioned above.

Before measuring efficiency, use them both to measure an identical load, and verify that the wattages are identical.

## EMI and Lightning

We offer a few quick words about measurement of EMI and lightning. There are a variety of home-built methods of measuring EMI. For example, people put a 1x oscilloscope probe on the input and run it into a spectrum analyzer to determine the conducted emissions. To measure radiated emissions, people make a one turn loop of wire (a sniffer) and attach it to an oscilloscope probe and run that into their spectrum analyzer. While these methods may have some value for determining relative improvement of performance, generically they just aren't accurate enough. When you're ready, pay the money and go to an EMI lab. There's really no substitute. Usually you can measure lightning there as well.

As to what to actually do at the EMI laboratory, preparation is the key to success. Whenever we go to the EMI laboratory, we take along a box of all sorts of different sizes and values of inductors and capacitors. We do a quick scan (a prescan), and if the measurements are out we try adding in or swapping out a capacitor or inductor. You can always find something that works. Then when you go back to your own lab, you can find a design that fits onto the board and uses those values.

Determination of lightning goes by much the same concepts. Go with three of four different-sized MOVs. Use the biggest one first. If that passes your lightning requirements, go to the next smaller size. Continue until the unit blows up—then you know that that last size was too small.

Oh, and always take along at least two identical units for your EMI testing. There's nothing more frustrating than getting there and finding out that the unit that worked perfectly that morning has stopped working that afternoon.

## ACCELERATED LIFE TESTS

We've already talked about LM-80, which gives guidance on how to measure LED lifetime. But LED lifetime is not necessarily the only factor governing product lifetime. If the ballast fails in a month, or the plastic yellows in a year, these other components rather than the LEDs will have determined the lifetime of the product.

Let's suppose that you have already built units, and verified that the temperature of the LEDs is such that their lifetime is going to be adequate for your product per their manufacturer's tests. If your lighting product is guaranteed for 10,000 h, you can't afford to wait 15 months to collect lifetime data. And if you're getting an Energy Star rating that requires a 3 year warranty, management is going to want to be very sure that there aren't any failures during that time. What can you do?

The thing to do is to recognize that the lifetime and warranty issues are usually the very last step in testing. The reality is that most products coming from the development cycle have plenty of problems that show up right away, for example, in the first 1000 h ("infant mortality"). You want to progressively test longer and longer periods, to avoid wasted time. The first step in testing your new product should be to identify and fix these, as they are relatively easy and quick to find.

To get started, just try running the product for a while. You don't need a fancy oven. Plug in a few units to the AC outlet (or 5 V USB port, or whatever), and let them run overnight. When you come the next morning, there may be one or two units that don't work anymore. Debugging the failures will immediately strengthen your design, without your having to set up complicated resources to do a full life test.

How many units should you use? Six or 10 is probably the right sort of number. You're not looking to collect statistically significant data; you just want to catch any significant design problems. In reality, this test may be done with hand-built units, rather than production units. That's okay, do it with just three. But make sure all three have identical components, so that you're catching a life issue, not still doing design work.

As a next step for lighting products, try a longer life test together with a cycle test. For example, get 10 units from a preproduction run and run them on the AC line for a week (this doesn't preclude you from continuing to run tests on the hand-built units). At the same time, switch on and off three units a thousand times. A cycle test such as this can catch start-up problems that a continuously on test won't see. You don't have to turn the switch on and off one thousand times (although this can be done, we've done it). For under $100 you can buy an AC line timer that switches power on and off at a fixed period. Two seconds on and then 2 s off is a period of 4 s, so in one day you can run 20,000 cycles!

After you've passed these preliminary tests you can move on to life tests. Of course, you still can't afford to wait 10,000 h. And, in fact, even 10,000 h of testing isn't long enough to give you good statistical data on the MTTF (meantime-to-failure) of your units. MTTF is what management really wants to know: Over the 10,000 h guaranteed life, how many units will actually fail?

In order to know how many failures there are going to be before a given time, you have to actually have some failures. Suppose, for example, that you test your product for 10,000 h and there are no failures. Perhaps if you had tested them for 10,100 h, 35% would have failed (the electrolytic capacitor wears out, etc.). The problem here is that to reliably estimate how many are going to fail in the 10,000 h, you need to know both the mean time to failure and the standard deviation around this mean. In this example, suppose the MTTF is something like 10,200 h, and the standard deviation is 100 h. That means that 10,000 h is only two standard deviations away from the mean. Then you should expect that in the first 10,000 h you will get exp $(-2) = 13.5\%$ of the units to fail—clearly unacceptable.

The goal is to accelerate the failure rate so you can collect the MTTF data. As already discussed in Chapter 5, most failures are governed by the Arrhenius law. They can thus be accelerated by increasing temperature. So what we want to do is to put the products into a thermal chamber and heat them past normal operating conditions to get them to fail prematurely.

If only it were that simple! Although temperature accelerates the failure rate, you don't know *how much* it accelerates it. So to use an accelerator, you also have to collect enough information to know how much acceleration a given temperature change results in. Of course, this information will be useful to you anyway.

Presumably the failure rate in Phoenix, Arizona, is going to be higher than that in Juneau, Alaska.

Here's what to do practically. You need two temperature-controlled ovens. Set one oven to run 40 °C hotter than the average ambient temperature in which the product is going to be set up. Set the second oven to be 30 °C hotter. Now into each oven place 30 units, and put a final 30 units in the ambient temperature in which the product will be used.

The hottest oven will presumably cause the units to fail first. Carefully record the time of each failure to the nearest hour. (This can be done, for example, with a recording power meter; when a unit fails, the power drops.) When half of the units have failed, that's the MTTF. And it's straightforward on Excel to calculate the standard deviation of the failure rate.

Now you must wait a while longer (usually about twice as long) for half of the units in the 30 °C oven to fail. You collect the data on these as well, and find the MTTF. Since you know the mean at both 40 and 30 °C above actual operation, you can compute what the mean will be in the actual environment by assuming an exponential relationship. For example, suppose the mean failure at 40 °C takes 600 h, and at 30 °C it took 1400 h. For each factor of 10 °C increase, the mean failure rate increases by $1400/600 = 2.33$. Therefore, at 20 °C above normal, the MTTF will be $1400\,h \times 2.33 = 3250\,h$, at 10 °C above normal the MTTF will be $1400\,h \times 2.33^2 = 7620$ h. And at normal operating temperature, the MTTF will be $1400\,h \times 2.33^3 = 17{,}800$ h. Since you also know the standard deviation of these numbers, you can now reliably estimate the failure rate at 10,000 h.

These instructions raised a few questions. Why did we pick 40 and 30 °C? The reason is that many components have roughly a "power of two" rule dependence on temperature: raising the temperature by 10 °C causes the MTTF to halve. So when we heard that we needed 10,000 h data, we estimated that 30 °C would accelerate the failure rate by $2^{(30°C/10°C)} = 2^3 = 8$, giving about 1200 h equivalent, 7 weeks. This seemed like a reasonable amount of time to wait. If the data were needed for 20,000 h, we would have picked 50 and 40 °C instead.

In that case, why not pick 50 and 40 °C right away? The problem is absolute maximum component temperatures. Some devices, such as capacitors, just have shorter lifetimes when overheated. This after all is the effect we're looking for. But some just fail right away when overheated. For example, if the plastic case melts, you're out of the temperature regime where the Arrhenius law applies. So you can't go above the maximum temperature at which catastrophic failure starts. This puts an upper bound on how much acceleration can be achieved. Beyond this, you just have to test the product for long enough, even if that means months of testing.

Why did we pick 30 units? Wouldn't 10 units be enough? The number has to do with the statistical certainty of the measurement. The uncertainty decreases as the square root of the number of units tested. One hundred units will give you $\sqrt{10} = 3$ times better accuracy than 10 units. As a compromise between using a large numbers of units and suffering from poor accuracy, 30 units has been generally agreed upon. You may be able to get by with fewer units, but you need to talk to an expert.

A few other issues should be mentioned. You don't really have to wait for half of the units to fail. You can get a number for the MTTF from just a few units' failure.

This will again affect the accuracy, and needs expert guidance. What if none of them fail in what ought to be a reasonable time? Good news! This provides you at least a lower bound on the failure rate, and if this is long enough, maybe you don't care how long the actual MTTF is. Of course, knowing what the MTTF actually is will provide more comfort to the finance department, which has to budget money for returns.

You might note that all of these calculations have assumed a Gaussian failure rate. One circumstance you might face where this is a faulty assumption is if there are two failure mechanisms with roughly the same rate. This will show up as a bimodal distribution of failures. An easy test for this is just to plot the failure time with a bar graph. If you can see that there are two lumps, this is diagnostic. You could also use Excel to look for nonzero skew. In either case, consult with an expert.

What were those 30 units running at ambient good for? Well, it's fine and good that we estimated the life time from accelerated tests. Chances are that the test has yielded a good number. But there's nothing like a real measurement. You should just keep running those 30 units, basically forever. After a year or two, you'll start to see failures. That will confirm your calculations. Even if you change the design as a result of your accelerated testing, you should consider leaving those 30 original design units running. Those units will always have the most hours on them, even when you start the next 30 new design units.

Finally, some types of failure aren't accelerated by temperature. An example is mechanical flexing. If your product has to be folded back and forth in normal usage, such as the USB light is, you need to make sure that the unit doesn't mechanically fail. There are special vibration tables built to test the strength of mechanical design. But for something like the USB light we have another suggestion. Give it to your child to play with. If it's still intact the next day, it's probably a reasonably good design.

# CHAPTER *15*

# PRACTICAL MODELING OF LEDs

In this final chapter, we take a look at building computer models of LEDs and of the thermal systems to which they are attached. Computer simulations can take a wide variety of forms. At their simplest, they might just be an Excel spreadsheet that calculates the operating point of an LED. At their most complex, they might be a multiphysics simulation that simultaneously models the electrical and optical performance of the LEDs, their driver, the thermal system in which they operate, and the dynamic interaction between all of these.

Practical models tend to fall between these two extremes. You want sufficient detail to get some design estimates, but you can't spend a month doing tedious calculations to set up the model, either. We'll be using Excel to find equations that fit our data, and Spice software to do simulations. Multiphysics-type simulations will be accomplished with Spice models emulating both thermal and optical performance of LEDs and their systems.

Why make a model at all? One reason is that there are so many LED choices out there; new ones appear every month. It wouldn't be practical to lay out a breadboard and test each one, not to mention the time and expense of getting samples. A model can quickly take information from the datasheet and accurately predict which LED is best for your application.

Models can be useful tools; the trick is to know when to use them. Some cases are so simple that simulation wouldn't be worthwhile; for example, nobody simulates an RC time constant. In other cases, simulation may not give reliable answers. An example of this is the performance of a MOSFET in a switch-mode power supply. Rise and fall times are very sensitive to the model parameters, and yet determine much of the power loss in the transistor. In this case, trying it out in the lab is the only reliable choice. Finally, some cases are sufficiently complicated that you can't guess the answer and lab tests take too much time. LEDs fall into this last category.

Many ICs these days have Spice models available on the Internet. Unfortunately, such is not the case for LEDs. In part this is doubtless due to their complexity. Unlike normal silicon ICs, they have not only electrical, but also thermal and optical characteristics, all of which play an important role in their performance. This chapter aims to show how to derive models from datasheet specifications. We'll also try to give some ideas about when a model is "good enough."

*Practical Lighting Design with LEDs*, Second Edition. Ron Lenk and Carol Lenk.
© 2017 by The Institute of Electrical and Electronics Engineers, Inc. Published 2017 by John Wiley & Sons, Inc.

## PRELIMINARIES

What is the benefit of computer simulation? The brief answer is that simulation is useful for things you can't easily measure in the lab. An example of this is production yield. You can build one or maybe even 10 devices in your lab. If all 10 work, then you could hope for a yield between 90 and 100%. But 10% fallout could wreck a business. And 90% yield is a bogus conclusion from the data, anyway. All 10 of your LEDs (and resistors, capacitors, etc.) came from the same lot. What happens when you get a new reel and the parts' values are different?

The appropriate method for determining yield is computer simulation. The two common methods are worst case analysis (WCA) and Monte Carlo (MC) simulations. WCA is the more conservative method for determining production yield. The method is detailed in the author's book on power supply design. Briefly, you vary the value of each component or parameter between its guaranteed maximum and minimum and see how that affects performance of the circuit to specification. It isn't easy in the lab to find an 80 nF capacitor and a 120 nF capacitor to determine worst-case performance with your 100 nF capacitor, but it's easy in your simulation. If some component causes the circuit to fall out of spec, you can either adjust the parameter or estimate how much yield loss it will produce.

The more common method of determining production yield is MC simulation. You again tell the computer the highest and lowest values of all the parameters, and it randomly picks values over that range for each. If you run this 1000 times, and only once fail specifications, you get a reasonable guess at the yield being 99.9%. (Note: This flat distribution with equal probability for any value in range may not hold for some devices. For example, LED efficacy seems to always be at the bottom of the bin. This is because manufacturers are cherry-picking devices. Make sure you watch out for this sort of nonflat distribution; it can affect yield dramatically.)

A related area in which simulation is useful is optimization. As part of the design process the circuit has to be optimized, which usually means meeting specifications at minimum cost. If you remove one LED from the circuit, the remaining ones have to run at higher current to produce the same light. But then they get hotter, which reduces the efficacy; the higher current also reduces the efficacy. Plus, the losses in your ballast may be higher. In cases like this, simulation is good because there are so many interacting factors. You need to be able to scan the parameters very widely to find the optimum. This is best done with a computer.

A few words are in line about the authors' preference in simulation tools. For simple calculations, we always use Excel. "Simple" here means that one line follows from the previous lines; there aren't any feedback loops. Excel is good, for example, for evaluating expected light output from an LED given its current and temperature.

For anything more complex, we use Spice software. This is in keeping with our strategy of keeping simulations simple and transparent (more on this later). There are very fancy simulators that are based on Spice and have large component libraries and cost thousands of dollars. But we prefer a very basic simulator with just the Spice primitives and a low cost. We always use AIM-Spice (AIM-Software, Trondheim, Norway; www.aimspice.com). A free version is offered that permits enough components for many applications; the full version is inexpensive. This is the simulator used for all the examples in this book. Its text-based input should be readily understandable and translatable to whatever simulator you are using.

Even the most basic Spice simulator contains fairly complicated models. For example, in AIM-Spice the transistor model Q has 41 specifiable parameters. The authors bet that not many people know what they all do—certainly we don't! As a result, some of these models we just don't use, unless truly "any device will do." So when it comes to modeling LEDs, we don't try to adjust the 18 parameters of the "D" diode model. We'll build a model from scratch, using a "B" source.

So what we're going to do, specifically, is to take datasheet parameters and curves, and fit functions to them. Once we have the function, we can write it as a "B" source, and this then is the model we'll be using. This curve-fitting approach requires some care and recognition of its limitations. For example, it is known that any five data points can be exactly fit by a fourth-order polynomial ($a_{o0} + a_1x + a_2x^2 + a_3x^3 + a_4x^4$). So if you pick five points on an $I-V$ curve, you can get exactly those values with this equation. But if you now try to find the value at a sixth point, you'll usually find that the equation's prediction is wildly off.

A better plan is to stick with the simplest equation possible. In practice, this is almost always a linear or quadratic equation, or an exponential one. It doesn't fit the curve exactly, but who says the graph was measured exactly? It will be good enough in the sense of least-squares.

Once you have this equation, it's usually trustworthy from the minimum to the maximum value for which you selected data. It doesn't go wildly out of control at in-between points. And it is usually a pretty reasonable guess for values somewhat beyond the limits—just remember that the validity at these values is unknown. You can't base a design on the hope that your equation is right beyond the limits for which it's been designed.

## PRACTICAL OVERVIEW OF SPICE MODELING

In this section, we're going to give a brief overview of how to use Spice. Of course, there are entire books written on the subject. But this is going to be more like a review of the main things you need to know, along with some practical tips on how to get things to run smoothly. The simulator used for the examples will be AIM-Spice.

Let's start by taking a look at the structure of an AIM-Spice listing. A simple model is shown in Table 15.1. It consists of a 1 mΩ resistor being fed by a 100 mA test source. The first line of the listing is a description of the model, in this case "RESISTOR." It would also be typical to use this as the file name.

The next line is the actual resistor model. A listing consists of the part reference number, its connections, and its values. Spice already has a model of a resistor, which

**TABLE 15.1   A Resistor Model**

```
RESISTOR
R1 IN 0 0.001
I1 IN 0 -0.1
```

is called "R." Since this is the first resistor in this model, the reference number is "R1." The next resistor, if there were one, would be "R2," and so on. The "R" part is necessary to tell Spice what model to use, but the numbering is arbitrary. The next part could just as well be "R100," although we recommend consecutive numbering, as three-digit numbers will be used for submodels below.

Next to be added to the list are the connections of the resistor "R1." Since a resistor has two terminals, it requires two connections, here "IN" and "0" (zero). The names of the connections are arbitrary, but should be selected to convey useful information. When you have a hundred connections, node 89 isn't going to remind you what it does. The node "0" is special; zero signifies ground.

Finally, the last element of "R1" is its value, in ohms. Here we choose 1 mΩ. You can also use common prefixes, such as "1 K" to signify 1000 Ω. But note that you don't type in "ohms"; the simulator already knows what the units are.

The third line in our "Resistor" model is a current sink, "I1." Note that the "I" model is a sink, not a source, so if current is to come out of it, it must have a negative value. Current sources have the same structure as resistors. "I1" is connected to "R1" at the node "IN" and has the value 100 mA.

The model is now complete. We expect to get a voltage of 100 µV on "IN," corresponding to 100 mA through a 1 mΩ resistor. To run the model, we push the "OP" button in the AIM-Spice window. As expected, $V(\text{IN}) = 100 \, \mu V$, the text signifying "the voltage at node IN equals 100 µV."

Since we've already introduced "I," we should mention some other capabilities of it and its voltage source equivalent, "V." Putting a number after either one's connections makes it a DC source. But there are other choices in addition. The most useful is "PULSE," which makes the source change value at some time and then change back later. You can consult the help section of AIM-Spice for details. The only thing to be careful of is not to make the initial value of "PULSE" different from the DC value. You might get something different in the transient simulation than in the operating point.

One more useful thing to do with an "I" or a "V" source is to sweep the value. This tells the simulator to repeatedly run a DC operating point on the circuit. Each sweep uses a different value of the source. You get to this operation by pushing the "DC" button on the AIM-Spice simulator. (Remember that the operating point is "OP.") You select the minimum and maximum values, as well as the step size, from the window that pops up.

To turn to other circuit elements, a capacitor is the same as a resistor, using "C" instead of "R." A typical listing would be "C1 IN 0 1N IC = 1." We have the part reference designator followed by the two connections (no polarity). Next is the value, in this case 1 nF. Finally, the "IC = 1" tells the simulator that the initial condition (voltage) on the capacitor is 1 V.

The next circuit element we need is "D," a diode model. As mentioned earlier, it has a huge number of values potentially to be set. But we don't need to. Practically, there are really only two uses for diodes in simulations. One use, obviously, is as an actual diode model. In this case we usually are not concerned with the exact details of the model; we just want it to conduct in one direction and not the other. We can use the model as is, without setting any parameters. Such a use is shown in Table 15.2.

**TABLE 15.2  A Diode, Including a Diode Model**

DIODE
D1 ANODE 0 DNORMAL
I1 ANODE 0 -0.1
.MODEL DNORMAL D

Unlike for "R" and "C," the diode model has a polarity. The anode connection is first, the cathode is second. The last item on the line, "DNORMAL," tells the simulator what model to use for the diode (in this case it is a fictitious model name). The last line of the listing is the model for the diode, ".MODEL DNORMAL D." First comes the word.MODEL (don't forget the period in front!). Next is the model name, here again the name we made up, DNORMAL. Then follows the name of the Spice model to be used, "D." Any nonstandard parameters would be listed last. Running the model, we see that the voltage at the anode is 774 mV at 100 mA.

The other use of "D" is when an ideal diode is needed, which means that it shouldn't have the forward drop of a normal diode. An ideal diode could be used, for example, when a voltage source is required that only sources and doesn't sink current, and we don't want the voltage-dependent drop across a normal diode. To make this, we can change the last line to ".MODEL DIDEAL D N = 0.01." With this model, the voltage at the anode is 7.7 mV.

As already mentioned, we won't use "D" for modeling an LED. With so many parameters available, there's no telling that you're getting the performance you want. Instead, we will directly build the performance we want using a "B" source. "B" is a "nonlinear dependent source." What this means practically is that, unlike the other models, you can write equations to govern the output voltage (or current). In AIM-Spice, the allowed functions are basic arithmetic, powers, trigonometric and inverse trigonometric, logarithmic and exponential, absolute value, and the Heaviside step function. This is more than enough to make any model needed.

One more model that will be used is "X." This is not properly a model in itself. It refers to a subcircuit that is defined at the end of the listing. It is useful when you want to have the same circuit appear in the listing multiple times. Rather than type in the details each time, you can call it "X" and just define it once. The details of usage will be explained later, when we use it to model LEDs.

Once in a while, you need to have a measurement of time in your Spice circuit. This can be easily done with a $C$–$I$ timer. Table 15.3 shows a circuit that generates 1 V/s, so that the voltage at "TIME" can be used as a surrogate for the time.

**TABLE 15.3  A Timer Model**

TIMER
I1 TIME 0 -1
C1 TIME 0 1

# WHAT NOT TO DO

After all these good things being said about simulations, are you ready to jump into the chapter and type the models here into your simulator? Not just yet! Simulations have pitfalls waiting to trap those who rush into them.

Let's start off with a rule for simulation models: *Simple models are best.* Many people (particularly simulation tool vendors) might disagree with this, but listen to this sad story. One of the authors was building a simulation model of a complex power system, the one aboard what is now called the International Space Station. The model worked fine, correctly showing the bandwidth and phase margin of the power converters. The strange thing was, telling the system to turn off didn't work. An hour's worth of viewing waveforms showed that the problem was that an LM319 comparator model didn't go low when its inputs told it to. Was the problem in the rest of the circuit or in the LM319 model? It didn't seem that the problem could be the IC model because it was provided as part of the simulation program. But extensive probing couldn't find anything relevant wrong with the rest of the circuit.

A special test simulation was run, with nothing but the LM319, a couple of function generators, and a pull-up resistor to a +15 V supply on the output. The pull-up resistor was there because the LM319 is an open collector comparator. This means that it can sink current but not source it, so it needs a pull-up resistor to a voltage to go high. The result was the same: The output was always high. Changing the pull-up resistor value didn't affect it. Changing the voltage supply didn't affect it. In frustration, the supply was removed entirely. The output was *still* high! In the end, the vendor who sold the simulator had modeled the LM319 as having a high output—the inputs didn't have any connection to the output.

Of course, they fixed the problem right away when I told them. But the message is that you're dependent on the people who modeled the devices in your simulation. You trust that they correctly modeled all of the devices you're using. You trust that they modeled the devices not only according to how they are normally operated, but also according to the operating region in which you're using them. (Did they include reverse breakdown in the diode?) When you upgrade from version 20.01 to 20.02, you're also hoping that the people who patched the software didn't break a model that was working before. Anyone who's used a computer knows how remote the chances are that the software is perfect.

As a result, we recommend using the most basic models possible. Our models rely exclusively on Spice primitives. Does this sound like taking a step back in time? It does, but the visibility into what you're actually getting is worth it. And hours wasted finding someone else's software bugs will convince you of this.

Many potential problems with computer simulation are avoided by staying simple, but not all. A problem that nobody likes to acknowledge is simple misspellings. If you type a "C" instead of a "D," the simulation will doubtless show strange things right away. But suppose instead of 10 kΩ you accidentally type 100 kΩ? Or 19 kΩ? It's easy to type an extra zero, or hit the nine instead of the zero. With a hundred lines of code, or a large schematic page, this sort of problem can go unnoticed for a long time. If it's a pull-up resistor, maybe it won't be a big deal, but what if it's part of a timing circuit?

A related problem is connecting components to the wrong nodes. Of course, this is particularly easy with a text editor, but it can happen with circuit schematics as well. Making clean schematics (no lines crossing each other unnecessarily) helps this somewhat. And you must make a schematic for any text editor anyway, with all the nodes and components labeled. We'll show an example later.

# WHAT TO DO

The solution to many of these problems is anchoring. You anchor a model by building it in the lab, and making sure that key parameters match your simulation. For example, suppose you are building a model of a current source running an LED. Go into the lab with a current source and measure the current with an ammeter, measure the temperature with a thermometer, and measure the light output with an integrating sphere. Then go to your model and set the current and temperature to the values you measured, and verify that the light output predicted by the model is correct (to some accuracy, perhaps 5%). If not, then fix the model—the data are right by definition! To reiterate: *Don't spend all your time on modeling. You have to go to the lab for a reality check.*

You need to do the same anchoring for each important output of your model. If your model is supposed to predict die temperature, then you need to program in just the current and verify that the temperature predicted matches the measured value. Most important, your measurement should be relatively close to the actual operating point of the circuit. Otherwise, your model might be working in a regime where something gives incorrect results and you won't know it. The best plan is to run several measurement/simulation pairs over a range of operating conditions. This enhances your confidence that the model will correctly predict operation. The one thing you don't want to do is to tweak the model so it matches the data at only the one point measured. Such models invariably give bad predictions at any other operating point.

We've already mentioned this in passing, but it's worth repeating. *You can't get out what you didn't put in.* As an easy example, suppose you've modeled an LED's I–V curve so that it correctly predicts forward voltage at a current of 350 mA. Will it give the right answer at the absolute maximum current of 1 A? Perhaps. Will it give the right answer at a peak pulse current of 14 A? No chance! You didn't program in things such as bond wire resistance or the formation of hot spots on the die. Chances are the LED will explode if you try it. That's why you need to anchor your model in the lab close to its real operating point.

# MODELING FORWARD VOLTAGE

Let's start by modeling the forward voltage of an LED as a function of current. Figure 15.1 shows the curve the vendor provides for this parameter, presumably from measured data. The data run from 100 mA to 1 A. Note that this curve is at 25 °C; the effects of temperature will be added later.

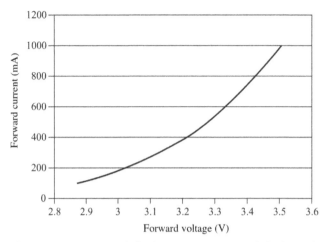

Figure 15.1 Luxeon Rebel I–V curve. (*Source:* Technical Datasheet DS56, Power Light Source Luxeon Rebel, Philips Lumileds Lighting Co., 2007.)

The curve looks fairly smooth. The scale of both the current and the voltage are linear, so a quadratic function might be a reasonable guess to fit the data. In this case, however, we know from engineering theory that a diode's voltage and current are exponentially related. So we're going to do a fit to $I = ae^{bV}$.

Let's start by collecting data points. The number of data points we collect isn't strictly determined. We should have more than three for reasonable accuracy and less than 10; it doesn't produce additional accuracy to have more. In practice, the number of points is usually determined by how many points are easy to read from the graph without having to interpolate.

We have entered the data in Excel in Table 15.4. We have selected 100 mA and 1 A because these are the endpoints of the graph. We picked 350 mA rather than 400 mA because the datasheet specifies forward voltage at that point. We'll use the specification later, but for now we just read the values from the graph. The other three points are chosen because they're spaced apart a bit and are easy to read. The point at 400 mA was skipped because it's so close to 350 mA.

**TABLE 15.4   Values from an *I–V* Curve for an LED**

|   | A | B |
|---|---|---|
|   | **I (mA)** | **V (V)** |
| 1 |  |  |
| 2 | 100 | 2.87 |
| 3 | 200 | 3.02 |
| 4 | 350 | 3.18 |
| 5 | 600 | 3.32 |
| 6 | 800 | 3.42 |
| 7 | 1000 | 3.50 |

**TABLE 15.5 LED *I–V* Curve with Logarithm of Current**

| | *A* | *B* | *C* |
|---|---|---|---|
| 1 | *I* (mA) | *V* (V) | ln(I) |
| 2 | 100 | 2.87 | 4.60517 |
| 3 | 200 | 3.02 | 5.298317 |
| 4 | 350 | 3.18 | 5.857933 |
| 5 | 600 | 3.32 | 6.39693 |
| 6 | 800 | 3.42 | 6.684612 |
| 7 | 1000 | 3.50 | 6.907755 |

Now we're going to fit the exponential function to this data. Excel has a function that does a least-squares fit to linear data, so we'll take the log of the current (because $I = ae^{bV}$ implies $\ln(I) = \ln(a) + bV$). In the next column, the formulas should read "=LN(A2)" and so on. Table 15.5 shows the result.

Now we can perform the linear least-squares fit. In cell B9 we use "SLOPE(C2:C7, B2:B7)," and in B10 we use "INTERCEPT(C2:C7, B2:B7)." See Table 15.6.

The equation we want is $I = \exp(\text{intercept} + \text{slope}^*\text{V})$. In D2 we type "=EXP(B$10+ B2*B$9)." The dollar signs are to keep the auto-incrementing of equations in Excel turned off for this value; we don't want the values sliding down when we paste the function into the other cells. We also check the error in E2 with "=(D2-A2)/D2." The final result, presented in Table 15.7, shows that we have fit the data points to better than 10% accuracy over the entire range. This is good enough for most practical purposes. We show all the formulas in Table 15.8.

So our model is $\ln(I) = -5.756 + 3.6398 * V$ with $I$ in mA. Notice that when $V = 0$, $I = 3\,\mu\text{A}$. Dark current isn't specified for this LED, but this number is practically zero for our purposes. Thus, the model is reasonable from 0 to 1 A. Note however, that the model is *not* reasonable for $V < 0$. The exponential is always positive, and so the model predicts that with reverse voltage applied to the LED,

**TABLE 15.6 Excel Fitting of LED *I–V* Curve**

| | *A* | *B* | *C* |
|---|---|---|---|
| 1 | *I* (mA) | *V* (V) | ln(I) |
| 2 | 100 | 2.87 | 4.60517 |
| 3 | 200 | 3.02 | 5.298317 |
| 4 | 350 | 3.18 | 5.857933 |
| 5 | 600 | 3.32 | 6.39693 |
| 6 | 800 | 3.42 | 6.684612 |
| 7 | 1000 | 3.50 | 6.907755 |
| 8 | | | |
| 9 | Slope | 3.6398 | |
| 10 | Intercept | −5.75564 | |

**TABLE 15.7 LED *I–V* Curve Model Showing Goodness of Fit of Equation**

| | *A* | *B* | *C* | *D* | *E* |
|---|---|---|---|---|---|
| 1 | *I* (mA) | *V* (V) | ln(I) | Eqn. | Error |
| 2 | 100 | 2.87 | 4.60517 | 108.91732 | 9% |
| 3 | 200 | 3.02 | 5.298317 | 188.02203 | −6% |
| 4 | 350 | 3.18 | 5.857933 | 336.61075 | −4% |
| 5 | 600 | 3.32 | 6.39693 | 560.31508 | −7% |
| 6 | 800 | 3.42 | 6.684612 | 806.31889 | 1% |
| 7 | 1000 | 3.50 | 6.907755 | 1078.8634 | 8% |
| 8 | | | | | |
| 9 | Slope | 3.6398 | | | |
| 10 | Intercept | −5.75564 | | | |

current is still positive! If you want to add in breakdown voltage, this has to be done separately. We'll talk about this below.

Now that we have the equation, we turn to Spice to create a model. Now, a "B" source can be either a current or a voltage source. Of course, we want to drive the LED with a current source. Two current sources in series would compete with each other in a simulation. So we use the voltage source form of "B," and thus the logarithmic form of the equation $V = 0.27474^{*}(\ln(I) + 5.756)$, with current in mA and voltage in volts.

Let's turn now to implementing the forward voltage equation in Spice. A listing is shown in Table 15.9. The model is of an LED being fed by a 100 mA current test source. The LED itself will be a compound device, consisting of a "B" source and a resistor. The "B" source is generating the LED's equation, while the resistor is a dummy in series with the "B" and is used to measure the current into the LED. We need to measure the current for the equation we're using for the LED.

**TABLE 15.8 Equations of Excel Model of LED *I–V* Curve**

| | *A* | *B* | *C* | *D* | *E* |
|---|---|---|---|---|---|
| 1 | *I* (mA) | *V* (V) | ln(I) | Eqn. | Error |
| 2 | 100 | 2.87 | =LN(A2) | =EXP(B$10 + B2*B$9) | =(D2-A2)/A2 |
| 3 | 200 | 3.02 | =LN(A3) | =EXP(B$10 + B3*B$9) | =(D3-A3)/A3 |
| 4 | 350 | 3.18 | =LN(A4) | =EXP(B$10 + B4*B$9) | =(D4-A4)/A4 |
| 5 | 600 | 3.32 | =LN(A5) | =EXP(B$10 + B5*B$9) | =(D5-A5)/A5 |
| 6 | 800 | 3.42 | =LN(A6) | =EXP(B$10 + B6*B$9) | =(D6-A6)/A6 |
| 7 | 1000 | 3.5 | =LN(A7) | =EXP(B$10 + B7*B$9) | =(D7-A7)/A7 |
| 8 | | | | =EXP(B$10 + B8*B$9) | |
| 9 | Slope | =SLOPE (C2:C7, B2:B7) | | | |
| 10 | Intercept | =INTERCEPT (C2:C7,B2:B7) | | | |

**TABLE 15.9   First Model of Luxeon Rebel**

LUXEON REBEL
B1 AN 0 V = 0.27474*(LN(V(ANODE)-V(AN)) + 19.5716)
R1 ANODE AN 0.001
I1 ANODE 0 −0.1

The LED consists of a node "ANODE" and has its cathode connected to ground. "ANODE" actually goes to the current-sense resistor. The resistor is then connected to the "B" source at an internal node "AN" (see Fig. 15.2; we will use the convention that node names are in boxes, reference designators are not). The "B" source goes from "AN" to ground. Its value is the equation $V = 0.27474*$(LN (V(ANODE)V(AN)) + 19.5716). Here, V(ANODE)-V(AN) is a measure of the current; it's the voltage across the resistor. The expression "V(ANODE)" means "use the value of the voltage at the node 'ANODE.'" Since this equation uses current in mA, we have used a 1 mΩ resistor. A value of 1 A through it will generate 1 mV, and so this is the correct value for the equation. We've multiplied the current in amps by 1000 to get the current in mA—that's why the resistor is 1 mΩ.

Now we can test our model. Pushing "OP" shows that the voltage at "ANODE" is 2.84675 V, comparing well with our graph value of 2.87 V. As expected, the voltage at "AN" is 100 μV lower than that at "ANODE." This small value doesn't matter practically, which is why we picked a small value resistor to measure the current in the first place.

We can now change the value of the current source to verify that the model generates the correct forward voltage. For example, at 1 A the model shows 3.48 V, compared with 3.50 V on the graph. In short, this model adequately reproduces the forward voltage of the LED. You might note that AIM-Spice also allows the current in "I1" to be swept, using the "DC" button. However, the plot is not very useful, because there's no straightforward way to plot current versus voltage, rather than the other way around.

Now consider what happens if we want seven LEDs in series. We could have just copied "B" and "R" seven times, but it isn't very efficient. It's better to build a model of the LED, to which we can then refer seven times. That's what we'll do next, as shown in Table 15.10.

In the model in Table 15.10, the LEDs are represented by the circuit model "X." "X" always refers to a subcircuit, the name of which is at the end of the line. Since

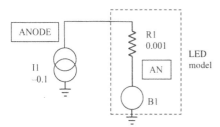

Figure 15.2   First model of Luxeon Rebel.

**TABLE 15.10  A Model of Multiple Luxeon Rebels in Series**

---

MULTIPLE LUXEON REBELS
X1 IN LED1 LED
X2 LED1 0 LED
I1 IN 0 -0.1
.SUBCKT LED ANODE CATHODE
B100 AN CATHODE V = 0.27474*(LN(V(ANODE)-V(AN)) + 19.5716)
R100 ANODE AN 0.001
.ENDS

---

LEDs have two connections, "X" does also. For other subcircuits, "X" could have any number of connections. "X1 IN LED1 LED" tells the simulator that the LED subcircuit is to be connected from nodes "IN" and "LED1" and is to use the model named "LED." Similarly, "X2 LED1 0 LED" tells it that the LED subcircuit is to be connected from "LED1" to ground and to use the same model, "LED."

The "X" model tells the simulator to look for a subcircuit called "LED." This is found in the last four lines of the model. The first of these says that it is a subcircuit called "LED" and that it has two connections, "ANODE" and "CATHODE." Note that there is a period in front of this line. The next two lines are the same as we used before. The only difference is that we have used reference designators beginning with 100 rather than 1, to signify that these models are inside a subcircuit. Finally, the subcircuit's last line is ".ENDS," again with a period in front. This statement is necessary to separate the subcircuit from (potentially) other subcircuits.

Running "OP" shows what we would expect. Node "LED1" is at the same 2.85 V that resulted before, and node "IN" is at twice that, 5.69 V. So now we have a model of an LED that can be used with a single line. We could now combine it with, for example, a model of a power supply. Instead, we will further improve our LED model. Next we will turn to reverse breakdown of the LED, before getting to optical output.

## REVERSE BREAKDOWN

The reader will have noticed that we have not addressed reverse breakdown voltage of the LED, except to note that the *I–V* model doesn't calculate *I* properly with negative *V*. There are two reasons for this. The first practical reason is that LED manufacturers recommend that you do not operate LEDs this way. Some of them don't even specify what the reverse breakdown voltage is. It's easy to break the LED by putting current through it in the wrong direction. Thus, for most applications, having a model for the reverse breakdown should be unnecessary.

A second reason is that the discussion about modeling reverse breakdown is going to get complicated. So the reader may skip this section, unless there is a particular need for a model to work with negative voltages. (The model we generate won't be used in the rest of the chapter.)

Those caveats out of the way, let's see what happens if we just apply a voltage from cathode to anode. The listing is shown in Table 15.11. By running a DC sweep

**TABLE 15.11   First Model of LED's Reverse Breakdown Characteristics**

```
LED REVERSE BREAKDOWN
X1 ANODE 0 LED
V1 ANODE 0 -5
.SUBCKT LED ANODE CATHODE
B100 AN CATHODE V = 0.27474*(LN(V(ANODE)-V(AN)) + 19.5716)
R100 ANODE AN 0.001
.ENDS
```

on this model, we can see that we get results that are not only implausible (the device probably doesn't conduct 11 kA at 5 V reverse) but also wrong: At 0 V the current should be 0 A, not −5000 A!

Let's fix the wrong result at 0 V first. This turns out to be a problem with the model, not with convergence. The LN function can't deal with zero voltage from "ANODE" to "AN," because the log of zero is negative infinity. Changing the value of V1 from −5 to 0 V and running the DC operating point shows what really is going on. The current is 3 µA. This isn't the real value; it's set by the GMIN option, minimum conductance. When the voltage on V1 is less than zero, the current is programmed to be even smaller, but it can't because of the GMIN. A little experimentation leads us to choose GMIN = 1.0E-20. Now the current at 0 V is 0 A, and the sweep shows that the current is zero from 0 V down to −5 V, to the resolution of the simulator.

Now that the current is zero in the reverse direction, we turn to making the device conduct at 5 V reverse. What we're going to do is fairly simple conceptually. In parallel with the model we've built so far, we're going to put a 5 V voltage source in backward. That way, when the voltage applied to the LED goes negative nothing happens until it becomes more negative than 5 V, and then it starts conducting (Figure 15.3).

We can't do literally this, because a voltage source will always try to source current. So in series with the voltage source we'll have to add a diode. This prevents the voltage source from sinking current when it's not supposed to, and additionally gives the benefit of a realistic *I–V* curve for the reverse breakdown. The model is shown in Table 15.12 and Figure 15.4.

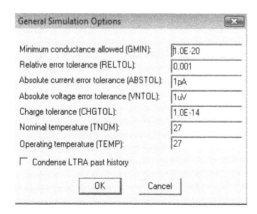

Figure 15.3   Setting the GMIN option to 1.0E-20 fixes the current at 0 V.

**TABLE 15.12 Improved Model of LED's Reverse Breakdown Characteristics**

---

LED REVERSE BREAKDOWN
X1 ANODE 0 LED
V1 ANODE 0 -5
.SUBCKT LED ANODE CATHODE
B100 AN CATHODE V = 0.27474*(LN(V(ANODE)-V(AN)) + 19.5716)
R100 ANODE AN 0.001
V100 CATHODE CLAMP 4.3
D100 CLAMP ANODE DMODEL
.ENDS
.MODEL DMODEL D

---

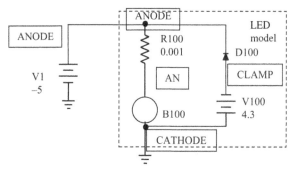

Figure 15.4   LED model including reverse breakdown.

We've set the "V100" source to 4.3 V rather than 5 V because the diode has a forward voltage of about 0.7 V. The diode is modeled as a "DMODEL," which we define at the bottom as just being the normal "D" model. This is done with the statement ".MODEL DMODEL D."

Now, when we run DC on this, we see that the current with 5 V reverse on the LED is 5 mA, and at 5.1 V it's up to 270 mA. Doing a DC sweep from −5.3 V to +3.2 V shows both the forward voltage and reverse breakdown characteristics expected, as shown in Figure 15.5.

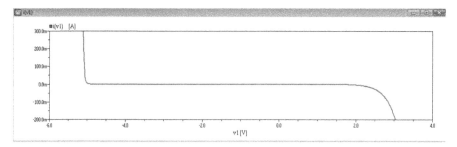

Figure 15.5   With the breakdown modeled, the complete *I–V* curve is shown. The *y*-axis is the negative of anode-to-cathode current.

# MODELING OPTICAL OUTPUT

We've built a model of an LED that correctly determines its forward voltage as a function of the current through it. As a next step, we want to determine the light output from the LED. Again, we're going to be working at 25°C, and will add thermal effects later. We're starting with the model shown in Table 15.10 (not including the breakdown effects of the last section).

Just as for forward voltage, we're going to fit some curves. The graph in Figure 15.6 shows (normalized) light output as a function of current. To obtain real lumens from this, we have to know exactly which device we're using. Let's assume that our part number produces 100 lumens at 350 mA. Thus, the graph's $y$-axis can be multiplied by 100 lumens to get light output. Our values extracted from the graph are shown in Table 15.13. We've added in 0 lumens at 0 mA to enhance the low current data.

It's tempting now to just do a linear fit of the equation, since the curve looks fairly straight. The only drawback is that at low currents, the light seems to drop off faster than it would linearly as current decreases. So we're going to model this as a quadratic, in order to better capture the low current light output.

For this purpose, we use an Excel graph, as shown in Figure 15.7.[1] We've graphed the data, and then added a trendline. We picked a polynomial trendline of order two, so that it's a quadratic fit, and checked the boxes to display the equation and the $R^2$ value. $R^2$ value is 0.999, indicating an outstandingly good fit.

As an aside, you might wonder why we didn't check the Excel box forcing the equation to be zero at zero. If you try this, you'll see that the correlation coefficient

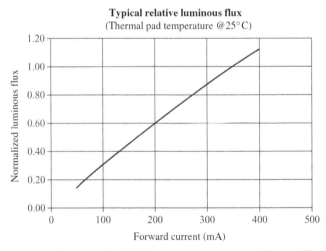

Figure 15.6   Normalized light output versus current. (*Source:* Technical Datasheet DS56, Power Light Source Luxeon Rebel, Philips Lumileds Lighting Co., 2007.)

---

[1] Note that this is a scatter plot, not a line plot. To get the trendline equation right, it *has* to be a scatter plot. Excel gives the wrong equation for a line plot!

**TABLE 15.13   LED Light Output as a Function of Drive Current**

|   | A | B |
|---|---|---|
| 1 | I (mA) | Light (Lm) |
| 2 | 0 | 0 |
| 3 | 100 | 37 |
| 4 | 200 | 63 |
| 5 | 350 | 100 |
| 6 | 600 | 150 |
| 7 | 800 | 185 |
| 8 | 1000 | 212 |

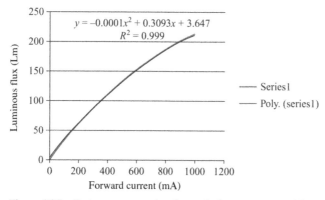

Figure 15.7   Data versus equation for optical output versus drive current.

actually gets worse (as does the error). This seems counterintuitive, until you realize that by removing the constant term, you've reduced the number of parameters from three to two. And fewer parameters make a worse fit, just as having enough parameters makes a perfect fit.

We can now apply the equation to our data, and verify that the error is less than 10% over the entire range (see Table 15.14). The error at 0 mA is of course a division by zero, but the error is actually only 4 lumens.

**TABLE 15.14   Spreadsheet Showing Goodness of Fit of Equation for LED Light Output as a Function of Drive Current**

|   | A | B | C | D |
|---|---|---|---|---|
| 1 | I (mA) | Light (Lm) | Eqn. | Error |
| 2 | 0 | 0 | 3.647 | #DIV/0! |
| 3 | 100 | 37 | 33.577 | −9% |
| 4 | 200 | 63 | 61.507 | −2% |
| 5 | 350 | 100 | 99.652 | 0% |
| 6 | 600 | 150 | 153.227 | 2% |
| 7 | 800 | 185 | 187.087 | 1% |
| 8 | 1000 | 212 | 212.947 | 0% |

**TABLE 15.15   A Model of Two LEDs Including Optical Output, at Fixed Temperature**

TWO REBELS WITH OPTICAL OUTPUT AT FIXED TEMP
X1 IN LED1 OPTIC1 LED
X2 LED1 0 OPTIC2 LED
I1 IN 0 -1
.SUBCKT LED ANODE CATHODE OPTICAL
B100 AN CATHODE V = 0.27474*(LN(V(ANODE)-V(AN)) + 19.5716)
B101 OPTICAL 0 V = −0.0001*((V(ANODE)-V(AN))*10^6)^2+0.3093*((V(ANODE)
   V(AN))*10^6) + 3.647
R100 ANODE AN 0.001
R1000 OPTICAL 0 1MEG
.ENDS

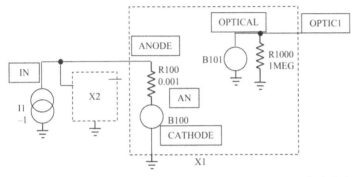

Figure 15.8   LED model with light output, temperature not yet included.

To include this equation in our Spice model is simple. Just as the voltage of the LED is a value at a node (the anode), the light output is going to be a value at a new node. We'll call it "OPTICAL." As shown in Table 15.15 and Figure 15.8, we add a "B101" source into the subcircuit with the equation in it. The voltage from "ANODE" to "AN" is 1 mV for 1 A, and we need it to be 1000, as that is the value in our equation, so we multiply by a million. The subcircuit now has another output, "OPTICAL," and it is listed as "OPTIC1" and "OPTIC2" in the main listing. The light output is obtained by looking at the voltage on these nodes. And indeed, at 1 A through the LED, the optic nodes show 213 V, which is to say 213 lumens, just as the graph shows. Note that inside the model we have added dummy resistors, "R1000," because the simulator does not recognize nodes that lack connections. Dummies are usually 1 MΩ, to keep the simulator from registering very small currents.

## MODELING TEMPERATURE EFFECTS

The LED model now accounts for both the forward voltage and the optical output as a function of current—but only at 25°C. Next we need to add in the effects of temperature. Spice has a global setting for the temperature. You can consider this

**TABLE 15.16   Model of a Luxeon Rebel Including Temperature Effect on Light Output**

MULTIPLE LEDS WITH OPTICAL OUTPUT AND TEMP EFFECTS
X1 IN LED1 OPTIC1 THERM1 AMB LED
X2 LED1 0 OPTIC2 THERM2 AMB LED
V1 AMB 0 27
R1000 AMB 0 1MEG
I1 IN 0 -1
.SUBCKT LED ANODE CATHODE OPTICAL THERM AMB
B100 AN CATHODE V = 0.27474*(LN(V(ANODE)-V(AN)) + 19.5716)-0.003*(V(THERM)
   V(AMB))
B101 OPTICAL 0 V = −0.0001*((V(ANODE)-V(AN))*10^6)^2+0.3093*((V(ANODE)
   V(AN))*10^6) + 3.647
R100 ANODE AN 0.001
R1000 OPTICAL 0 1MEG
R1001 THERM 0 1MEG
V100 THERM 0 60
.ENDS

to be the ambient temperature. Since the LEDs will be dissipating power, they will be hotter than the ambient temperature, and thus need to have their own temperature parameter. We'll call this node "THERM."

The effect of temperature on forward voltage is straightforward. The only data provided is that it is −3 mV/°C. What we're going to do for our model is create a voltage "AMB" corresponding numerically to the ambient temperature. This is done with a "V" source. This is then passed in to the LED model on one of its nodes.

Internal to the LED, we create a "THERM" node. For the moment, this is going to just be a settable value, using "V100." When we build a thermal model for the LED,

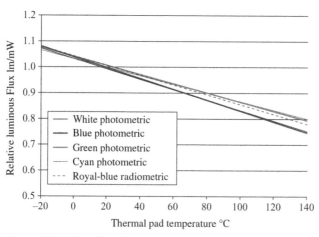

Figure 15.9   The effect of temperature on light output. (*Source:* Technical Datasheet DS56, Power Light Source Luxeon Rebel, Philips Lumileds Lighting Co., 2007.)

**TABLE 15.17 A Model of Two Rebels Including Temperature Effects on Forward Voltage**

TWO REBELS WITH TEMPERATURE EFFECT ON FORWARD VOLTAGE
X1 IN LED1 OPTIC1 THERM1 AMB LED
X2 LED1 0 OPTIC2 THERM2 AMB LED
V1 AMB 0 27
R1000 AMB 0 1MEG
I1 IN 0 -1
.SUBCKT LED ANODE CATHODE OPTICAL THERM AMB
B100 AN CATHODE V = 0.27474*(LN(V(ANODE)-V(AN)) + 19.5716)-0.003*(V(THERM)
    V(AMB))
B101 OPTICAL 0 V = (−0.0001*((V(ANODE)-V(AN))*10^6)^2 + 0.3093*((V(ANODE)
    V(AN))*10^6) + 3.647)* (1-0.0021*(V(THERM)-20))
R100 ANODE AN 0.001
R1000 OPTICAL 0 1MEG
R1001 THERM 0 1MEG
V100 THERM 0 60
.ENDS

it will be determined by the thermal contact of the LED to the ambient temperature. Finally, we need to add dummy resistors as before to ensure convergence. The model including thermal effects on the forward voltage is shown in Table 15.16. Running it as shown shows that the forward voltage at "ANODE" has dropped 100 mV, as expected for a 33 °C temperature rise, 33 °C × −3 mV/°C = −99 mV.

Finally, we need to add the effect of temperature on the light output. Looking at Figure 15.9, we see that this too is linear. At 20 °C the factor is 1.00, while at 140 °C it is 0.75. The slope is thus (0.75 − 1.00)/(140 °C − 20 °C) = −0.0021/°C, a decrease in light of about 1% for every 5 °C increase in temperature.

The model including this is shown in Table 15.17, with the schematic shown in Figure 15.10. We have multiplied the normal light output by the factor above times the temperature of the LED minus 20 °C.

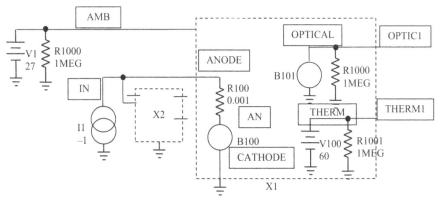

Figure 15.10   LED model with temperature effect on both forward voltage and optical output.

# MODELING THE THERMAL ENVIRONMENT

We've built a model of an LED, which was the main goal of this chapter. At the moment, however, this model has the LED temperature as an internally set parameter. What actually happens physically, of course, is that the power dissipated by the LED sets the LED temperature (along with the ambient). Our goal in this section is to model this.

To complete our LED model, we will add a thermal resistance from the device to the ambient temperature. This will replace the fixed temperature at "THERM with the calculated value based on the power dissipation of the LED. The power of course the current through the LED (as measured by "R100") multiplied by the voltage across the LED. We choose one of the thermal resistances to be 10 °C/W and the other 20 °C/W. Note that this thermal resistance is not in the LED mode itself. This is because it isn't a property of the LED. Since it is part of the external world, for example, the same as the drive circuit, it is included in the mail listing.

We add in another "B" in the model that calculates the power dissipated in the LED as the current through it multiplied by the voltage across it. This is the expression (V(ANODE)-V(AN))*10^3)*(V(ANODE)-V(CATHODE)), with the first term being the current. Since we now have electrical paths to ground for the "AMB" and "THERM" nodes, we can remove their dummy resistors. "OPTICAL" is the only node that still needs a dummy.

The completed LED model is shown in Table 15.18 with the schematic shown in Figure 15.11. Running it, we see that the one LED is at a temperature of 61 °C with a light output of 195 lumens, while the other is at 93 °C with a correspondingly reduced light output of 180 lumens. This model can now be attached to electrical components, for example, those modeling the power supply. But this is beyond the scope of this book.

**TABLE 15.18   Complete LED Model**

MULTIPLE LUXEON REBELS
X1 IN LED1 OPTIC1 THERM1 AMB LED
X2 LED1 0 OPTIC2 THERM2 AMB LED
V1 AMB 0 27
R1 THERM1 AMB 10
R2 THERM2 AMB 20
I1 IN 0 -1
.SUBCKT LED ANODE CATHODE OPTICAL THERM AMB
B100 AN CATHODE V = 0.27474*(LN(V(ANODE)-V(AN)) + 19.5716)-0.003*(V(THERM)
    V(AMB))
B101 OPTICAL 0 V = (−0.0001*((V(ANODE)-V(AN))*10^6)^2 + 0.3093*((V(ANODE)
    V(AN))*10^6) + 3.647)*(1-0.0021*(V(THERM)-20))
B102 THERM 0 I = −((V(ANODE)-V(AN))*10^3)*(V(ANODE)-V(CATHODE))
R100 ANODE AN 0.001
R1000 OPTICAL 0 1MEG
.ENDS

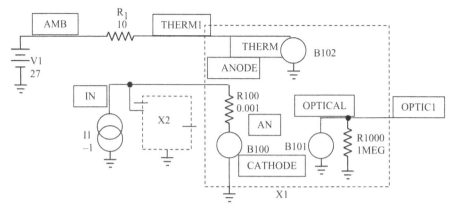

Figure 15.11   Complete LED model.

# A THERMAL TRANSIENT

As an application of this LED model, let's take a look at a thermal transient. We're going to step the current through the two LEDs from 100 mA to 1 A, and add two different thermal capacitances to the model so they respond with different speeds. The Spice listing is shown in Table 15.19. The thermal response in Figure 15.12 shows that the LED with the shorter thermal time constant reaches its new steady state temperature in about 5 s. The other LED takes nearly 20 s. The shorter time constant LED also has lower thermal resistance to the ambient temperature, and so doesn't get as hot (50 °C versus 80 °C).The optical output shown in Figure 15.13 shows that

**TABLE 15.19   Spice Model of Multi-LED Thermal Transient**

THERMAL TRANSIENT
X1 IN LED1 OPTIC1 THERM1 AMB LED
X2 LED1 0 OPTIC2 THERM2 AMB LED
V1 AMB 0 20
R1 THERM1 AMB 10
R2 THERM2 AMB 20
C1 THERM1 AMB 0.1 IC = 20
C2 THERM2 AMB 0.2 IC = 20
I1 IN 0 -0.1 PULSE(-0.1 -1 1U 1U 1U 25 100)
.SUBCKT LED ANODE CATHODE OPTICAL THERM AMB
B100 AN CATHODE V = 0.27474*(LN(V(ANODE)-V(AN)) + 19.5716)-0.003*(V(THERM)
    V(AMB)) ^^
B101 OPTICAL 0 V = (−0.0001*((V(ANODE)-V(AN))*106)2 + 0.3093*((V(ANODE)
    V(AN))*10^6) + 3.647)*(1-0.0021*(V(THERM)-20))
B102 THERM 0 I = −((V(ANODE)-V(AN))*10^3)*(V(ANODE)-V(CATHODE))
R100 ANODE AN 0.001
R1000 OPTICAL 0 1MEG
.ENDS

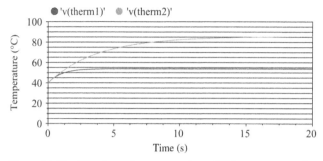

Figure 15.12    The thermal response of two LEDs to current pulse.

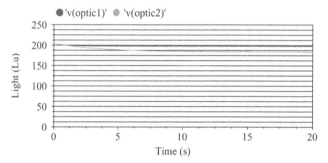

Figure 15.13    The optical response of two LEDs to current pulse.

corresponding to the hotter temperature, the longer time constant LED puts out less light (185 lumens versus 200 lumens).

## SOME COMMENTS ON MODELING

The foregoing sections epitomize our approach to modeling. You build up one section at a time, and check that it works before adding the next piece. Trying to debug it all at once is too difficult. Of course, the same thing could have been done with Spice with a schematic entry system. It's just one extra interface with potential problems of which you have to be aware.

You might notice that none of the models developed had any convergence issues. The Spice software is actually fairly robust. Convergence problems tend to arise only when there are abrupt transitions. An example would be using an ideal switch, "SW," for a transistor. In this case, minimizing the stiffness of the problem is usually helpful. The stiffness is the ratio of the biggest number to the smallest number. For example, with the switch, the ratio of the off-resistance to the on-resistance is the stiffness to be minimized. If 1 Ω is a good model for the transistor, don't model the switch as 1 mΩ. There are global parameters in Spice for helping along convergence, such as RELTOL, but our general feeling is that it shouldn't be necessary to tinker with these. If you're having convergence problems, you should work on your model further.

# REFERENCES

Alliance for Solid-State Illumination Systems and Technologies (ASSIST) (2006) ASSIST recommends . . . LED life for general lighting: recommendations for the definition and specification of useful life for light-emitting diode light sources. 1(7), February 2005, revised April 2006.

Betten, J. and Kollmant, R. (2007) LED lighting illuminates buck regulator design. *Power Electronics Technology*, October, p. 38. Available at http://powerelectronics.com/mag/710PET24.pdf.

Dowling, K. (2009) LED lighting standards and guidelines are now building on a firm foundation. *LEDs Magazine*, May/June.

Haitz, R. (Editorial Board) (2007) Editorial on Haitz's law, *Nature Photonics*, 1 (23): pp. 23–4.

Jameson, K.A., Susan, M.H., and Wasserman, L.M. (2001) Richer color experience in observers with multiple photopigment opsin genes. *Psychonomic Bulletin & Review*, 9 (2): 244–261.

Kalloniatis, M. and Charles L. (2007) *Principles of Vision*. Webvision. http://www.ncbi.nlm.nih.gov/bookshelf/br.fcgi?book=webvision&part=ch24psych1.

Lenk, R. (1998) *Practical Design of Power Supplies*. Hoboken, N.J.: IEEE Press/Wiley.

MacAdam, D.L. (1942) Visual sensitivities to color differences in daylight. *Journal of the Optical Society of America*, 32: pp. 247–74.

Martzloff, F. (1991) A standard for the 90s: IEEE C62.41 surges ahead. Available at http://www.eeel. nist.gov/817/pubs/spd-anthology/files/Standard%20for%2090s.pdf.

Ohno, Y. (2004) Color rendering and luminous efficacy of white LED spectra. *Proceedings of SPIE*, 5530: 88.

Scheidt, P. (2010) Innovative packaging improves LEDs' light output, lifetime and reliability. *Electronic Design News*, January.

Straka, T. (2009) Navigating the product safety certification process for solid-state lighting products. *LEDs Magazine*, November/December, pp. 37–38.

Tsao, J.Y., et al. (2010) Solid-state lighting: an energy-economics perspective. *Journal of Physics D: Applied Physics*, 43: 354001.

Turner, M. (1996) Available at http://www.electronics-cooling.com/articles/1996/may/may96_01.php.

Wong, B. and Zheng, L. (2009) Intelligent LED lighting systems and network technology choices. In: *Designing with LEDs*, e-book. *Electronic Design News*, April. Available at http://www.edn.com/article/459201-Designing_with_LEDs_E_Book.php.

Zong, Y. and Ohno, Y. (2008) New practical method for measurement of high-power LEDs. In: *CIE Expert Symposium 2008 on Advances in Photometry and Colorimetry*, pp. 102–106.

*Practical Lighting Design with LEDs*, Second Edition. Ron Lenk and Carol Lenk.
© 2017 by The Institute of Electrical and Electronics Engineers, Inc. Published 2017 by John Wiley & Sons, Inc.

# INDEX

*Practical Lighting Design with LEDs*, Second Edition. Ron Lenk and Carol Lenk.
© 2017 by The Institute of Electrical and Electronics Engineers, Inc. Published 2017 by John Wiley & Sons, Inc.

# IEEE Press Series on Power Engineering

Series Editor: **M. E. El-Hawary**, Dalhousie University, Halifax, Nova Scotia, Canada

The mission of IEEE Press Series on Power Engineering is to publish leading-edge books that cover the broad spectrum of current and forward-looking technologies in this fast-moving area. The series attracts highly acclaimed authors from industry/academia to provide accessible coverage of current and emerging topics in power engineering and allied fields. Our target audience includes the power engineering professional who is interested in enhancing their knowledge and perspective in their areas of interest.